FUNDAMENTAL PLACE-NAME GEOGRAPHY

Sixth Edition

ROBERT FUSON

University of South Florida

wcb

WM. C. BROWN PUBLISHERS
DUBUQUE, IOWA

Cover design by Jeanne M. Regan

Library of Congress Catalog Card Number: 88-70482

ISBN 0-697-05208-7

Printed in the United States of America by Wm. C. Brown Publishers
2460 Kerper Boulevard, Dubuque, IA 52001

10 9 8 7 6 5 4 3 2

To
Thomas F. Barton

Contents

Preface

In 1966, when the first edition of *Fundamental Place-Name Geography* appeared, scant attention was being given to the subject in the classroom. If anything, special efforts were made to avoid the teaching of locations. This partly came about as a result of the quantitative revolution, and what had been a science of place became a science of numbers.

In the last few years a number of people, from academicians to United States senators, have suddenly discovered that Americans are place-name illiterates. Survey after survey has confirmed this, and one leading syndicated columnist even suggested that a place-name quiz should be made part of the examination for a driver's license!

Even with the recent attention given to place-names, students of geography will probably continue to be the only reasonably informed group that can locate them on a map. And, more than likely, they will be among the few that comprehend the significance of the locations. This will have to come from the study of some appropriate subject matter. Learning the names of every place on earth is not geography; it is a means to an end, not the end in itself. Location is to geography as date is to history or the alphabet is to written language. It is a departure point. And it is a subject that is in constant flux.

Since 1966 the number of place-name changes has been phenomenal. *Forty* new, independent countries have been added to the world map, and more than *twenty-five* countries have changed names. Several political units have been absorbed by their neighbors, suggesting that geography's loss is history's gain. Many cities, rivers, and a variety of other features have lost their earlier designations. Chinese nomenclature now adheres to a system of romanization different from that of the 1960s, and it is only a matter of time before time-worn names such as Peking, Canton, and Tibet fade into obscurity. There is also a growing tendency to use local spellings for places within countries.

Generally, the greatest number of changes has occurred among the developing nations, reflecting the dissolution of the European colonial system and the disposition of certain territories that were under United Nations trusteeships.

Africa has given birth to twelve new countries since the first edition of this book was published. Namibia should soon become the thirteenth. The independence of Zimbabwe in 1980 marked the end of European political control on that continent.

Three regions have surpassed Africa in the percentage change of sovereign states. Excluding large, peripheral, traditional states, the Pacific region possessed only one independent nation in 1966; today there are nine. The Caribbean went from five to thirteen, and three new states now occupy the Indian Ocean, which used to have only one. South America has gained two additional countries; Southern and Southeast Asia gained two. Only Europe, Australia/New Zealand, and the U.S./Canadian/Mexican portion of North America have remained territorially stable in recent years.

Other changes in the world have kept apace. Population, for instance, has risen sharply. There are 58,000,000 more Russians today than there were at the time of the first edition, and there are 62,000,000 more Brazilians. The People's Republic of China has increased by about *three times the combined Russian/Brazilian increase.* This is a number larger than the total population of either the United States or the USSR!

These population data are updated in the Appendix. Data for each of the world's 170 sovereign states, the principal overseas territories of major countries, and the first-order political subdivisions of Canada, the United Kingdom, the United States, and the USSR are provided.

New maps of Japan and Israel have been added to the sixth edition, several maps have been re-drafted, and a pronunciation guide is furnished for selected place-names, where deemed appropriate.

A number of features from earlier editions have been retained. Metric equivalents of the standard English system of measurement and alternate spellings (usually local names) are two such examples. Outline maps for practice and testing are included, as before.

Despite extensive revision, and even expansion in places, the overall length of the text remains about the same as that of previous editions. The book should continue to be regarded as a supplement to a standard geographical text. It is neither a condensed world geography nor is it an atlas. The latter, in fact, should always be at hand.

It is hoped that this brief mental excursion to some of the world's named places will whet the appetite, not only to discover more locations but to know more about them.

ROBERT H. FUSON, Ph.D.
Professor Emeritus of Geography

University of South Florida
Tampa, Florida

Acknowledgments

All cartograms and maps are based on projections copyrighted by McKnight and McKnight, Bloomington, Illinois, to whom full acknowledgment is given.

I would like to thank a number of people who have helped make this edition possible. First, my appreciation goes to H. Karen Kincheloe, North Harris County College; Gene C. Wilken, Colorado State University; and Donald A. Cope, Columbus College. They reviewed the last edition and the manuscript of this edition at several different stages. Their thoughtful critiques prompted many changes.

A special word of thanks goes to all of the students of GR 100 (*Introduction to Geography*) at Colorado State University who participated in the review process. I trust I have incorporated many of their comments in a satisfactory manner. Student input in a textbook is long overdue, and I am confident it will add a great deal to the final product.

1

The Continents and Oceans

Although geography as a discipline is concerned primarily with the land and water surfaces of the earth, most geographers direct their energies to that 30 percent of the planet comprised by land. The *six* great continental masses account for almost 99 percent of the world's total land area. Approximately 88 percent of the world's total population of slightly over 5,000,000,000 occupies five of the six continents (Antarctica has no permanent population). The remaining 12 percent of the world's people live on islands, which account for about 1 percent of the total land surface.

Continent	*km²* *(1000s)*	*mi²* *(1000s)*	*% of Total Land*	*Population 1988 (Millions)*	*% of Total Pop.*
1. Eurasia	53,906	20,733	36.0	3,100	62.0
2. Africa	29,916	11,506	20.0	600	12.0
3. North America	24,414	9,390	16.3	380	7.6
4. South America	17,667	6,795	11.8	280	5.6
5. Antarctica	14,300	5,500	9.6	nil	0.0
6. Australia	7,717	2,968	5.2	16	0.3
Total	147,920	56,892	98.9	4,376	87.5

The 70 percent of our planet covered with water consists of numerous seas, gulfs, bays, lakes, and *four* great oceans.

Ocean	*Area*	
	km²	*mi²*
1. Pacific Ocean	166,884,380	64,186,300
2. Atlantic Ocean	86,892,000	33,420,000
3. Indian Ocean	73,711,300	28,350,500
4. Arctic Ocean	9,521,720	3,662,200

Note: There is *no* Antarctic Ocean.

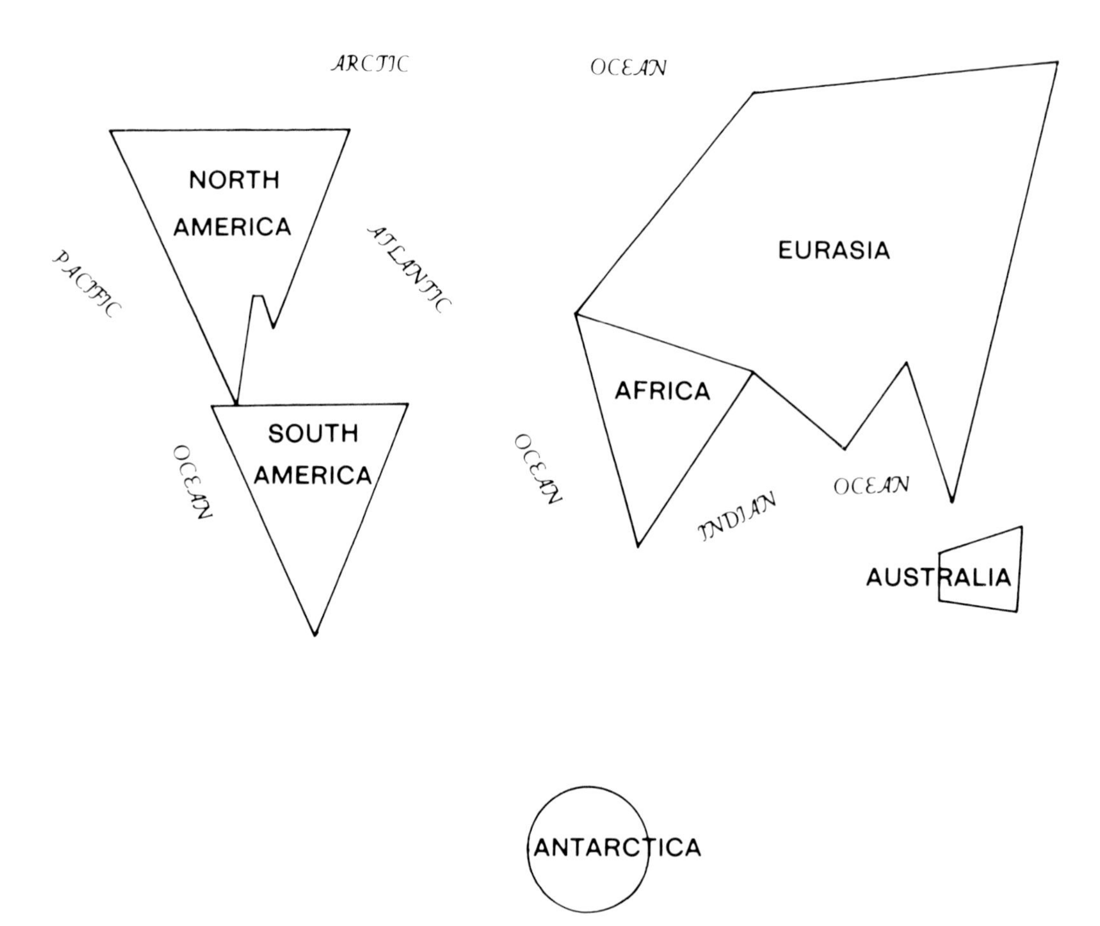

The relationship of the *six* continents and *four* oceans is indicated on the cartogram above.

Self-Test

Name the six continents.

____

____

____

Name the four oceans.

____

____

What percentage of the earth's land area is represented by the continents?

____

What percentage of the earth's surface is water? ______________

What percentage of the earth's population lives on continents? ______________________

On islands? ______________________

Approximately, what is the world's population? ______________________
Check your answers with page 1.
Be sure to check spelling. (Are there *two* letter c's each in Arctic and Antarctic?)
Continue *only* if the above answers are correct!

DO NOT LOOK BACK AT THE CARTOGRAM. DO NOT USE THE ATLAS. Label this cartogram exactly like the one studied on page 2.

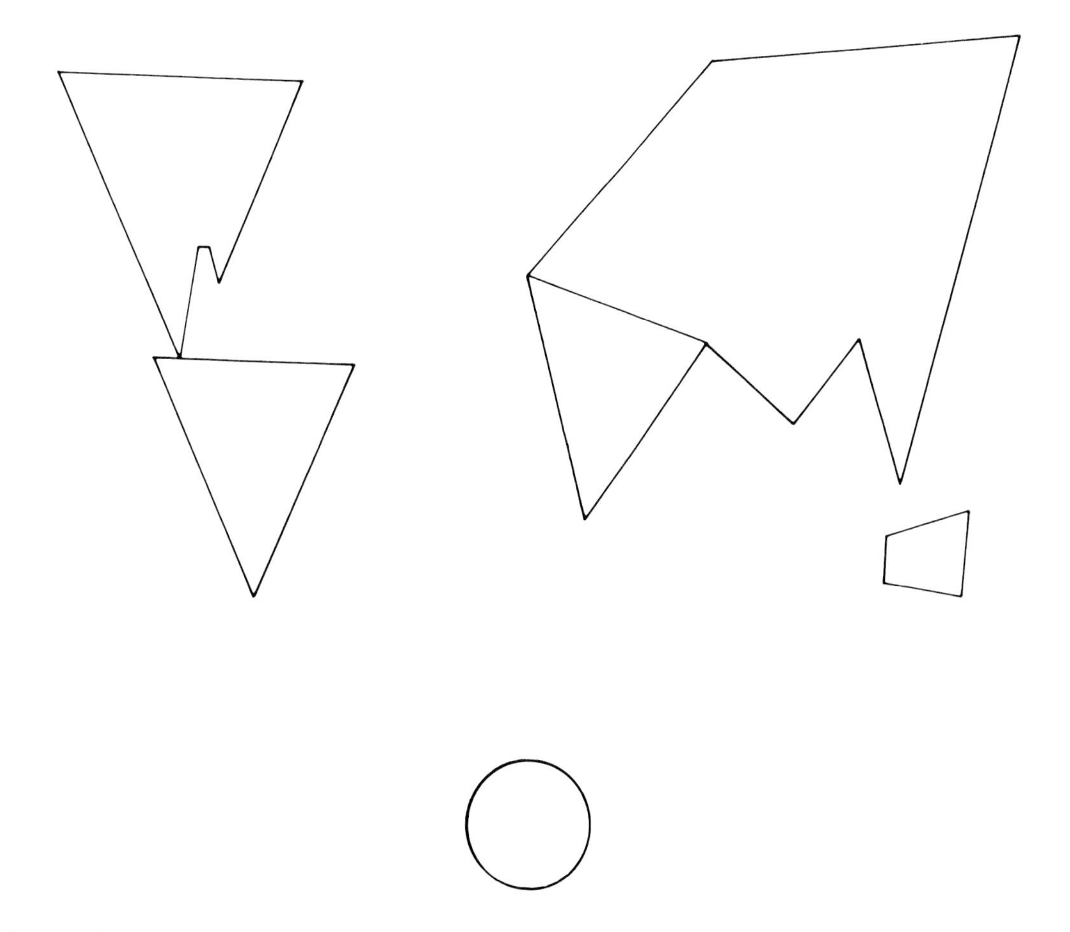

Check this work with page 2.
Correct any errors before continuing.

A WORD ABOUT TERRITORIAL CLAIMS IN THE ANTARCTIC

Seven states (Argentina, Australia, Chile, France, New Zealand, Norway, and the United Kingdom) claim portions of the Antarctic continent. British, Chilean, and Argentinian claims overlap and a large portion (between 90° W and 180°) remains unclaimed.

In 1959, the seven states above, plus Belgium, Japan, South Africa, the USSR, and the United States, signed the Antarctic Treaty. While the original claims were not voided, the treaty asserts that there shall be no new claims, that the continent shall be used for peaceful purposes only, scientific information shall be freely exchanged, and all installations are open to all for inspection.

Six additional countries have become consultative parties and another sixteen countries are non-voting parties to the treaty.

The treaty will be reviewed in 1991. Neither the United States nor the Soviet Union claims any territory in the Antarctic, and neither state recognizes any of the seven original claims.
The cartogram below illustrates a view looking straight down on the North Pole from outer space. Label the *four* oceans and the *five* continents.

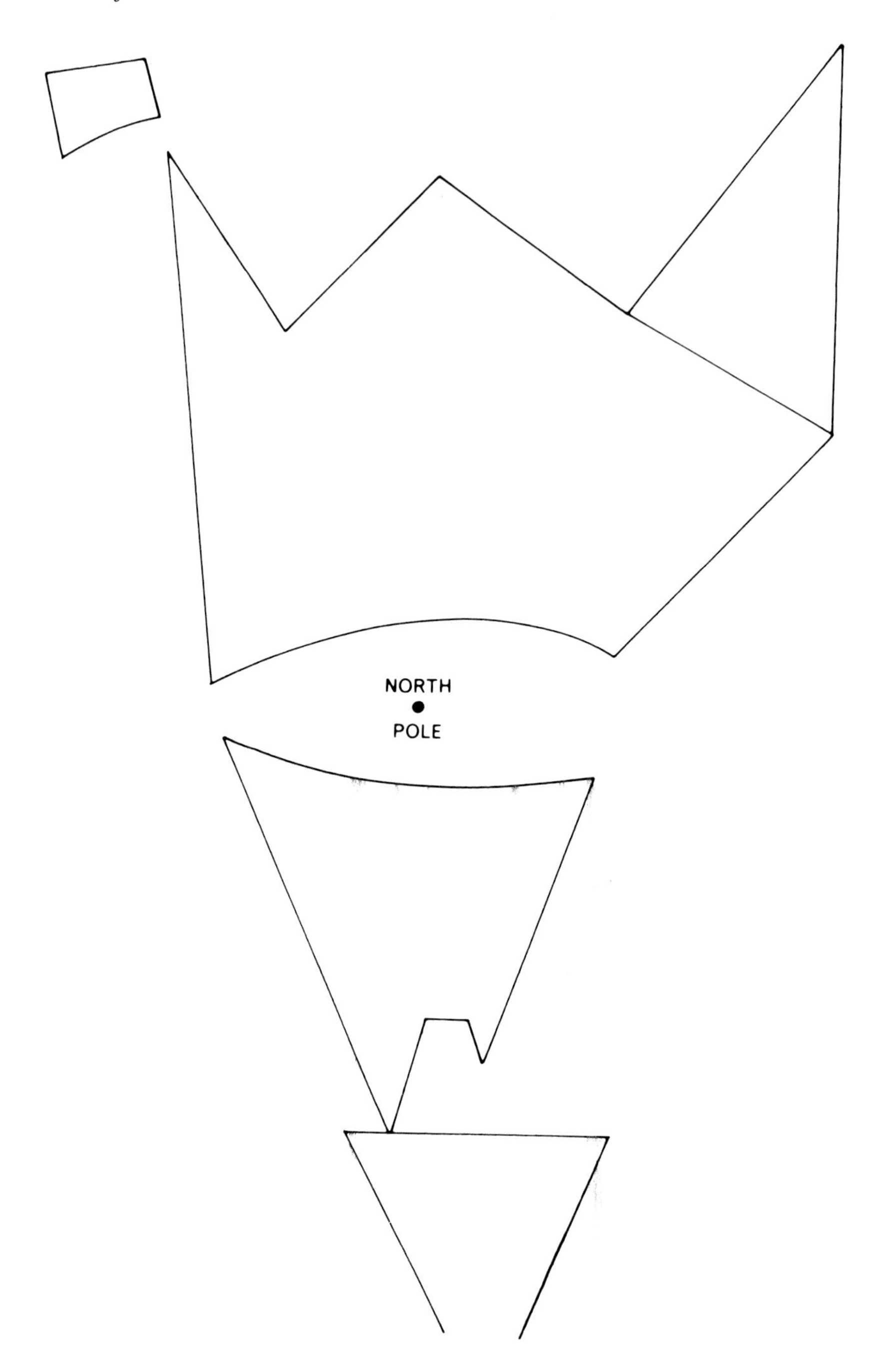

Check this work with page 2.
Correct any errors before continuing.

This is a view looking at the South Pole from outer space.
Label the *three* oceans and the *six* continents.

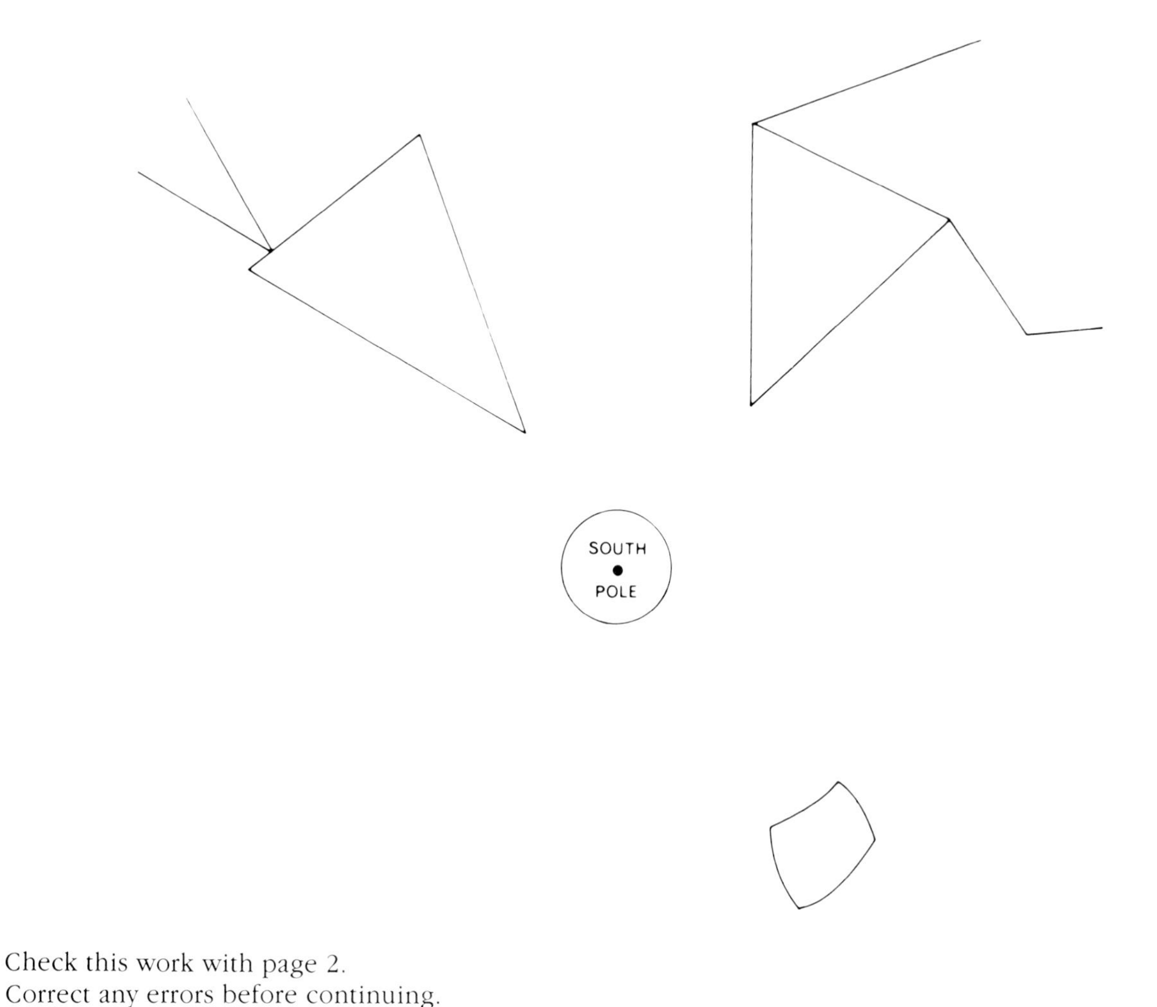

Check this work with page 2.
Correct any errors before continuing.

Europe is *not* a true continent, but merely a collection of peninsulas and islands in western Eurasia. It may, however, be located by a traditional boundary, as follows. Beginning in the north, the boundary between Europe and Asia follows the *Ural Mountains* southward to the man-made political boundary between the *Russian Soviet Federated Socialist Republic* (RSFSR) and the *Kazakh Soviet Socialist Republic* (see p. 60). This internal Soviet political boundary is followed to the *Caspian Sea,* thence along the south side of the *Caucasus Mountains* to the *Black Sea.* The Black Sea opens to the Mediterranean Sea, which separates Europe and Africa. On the cartogram on the next page, place every location that is italicized above. Then refer to the atlas and locate each feature labeled on the cartogram.

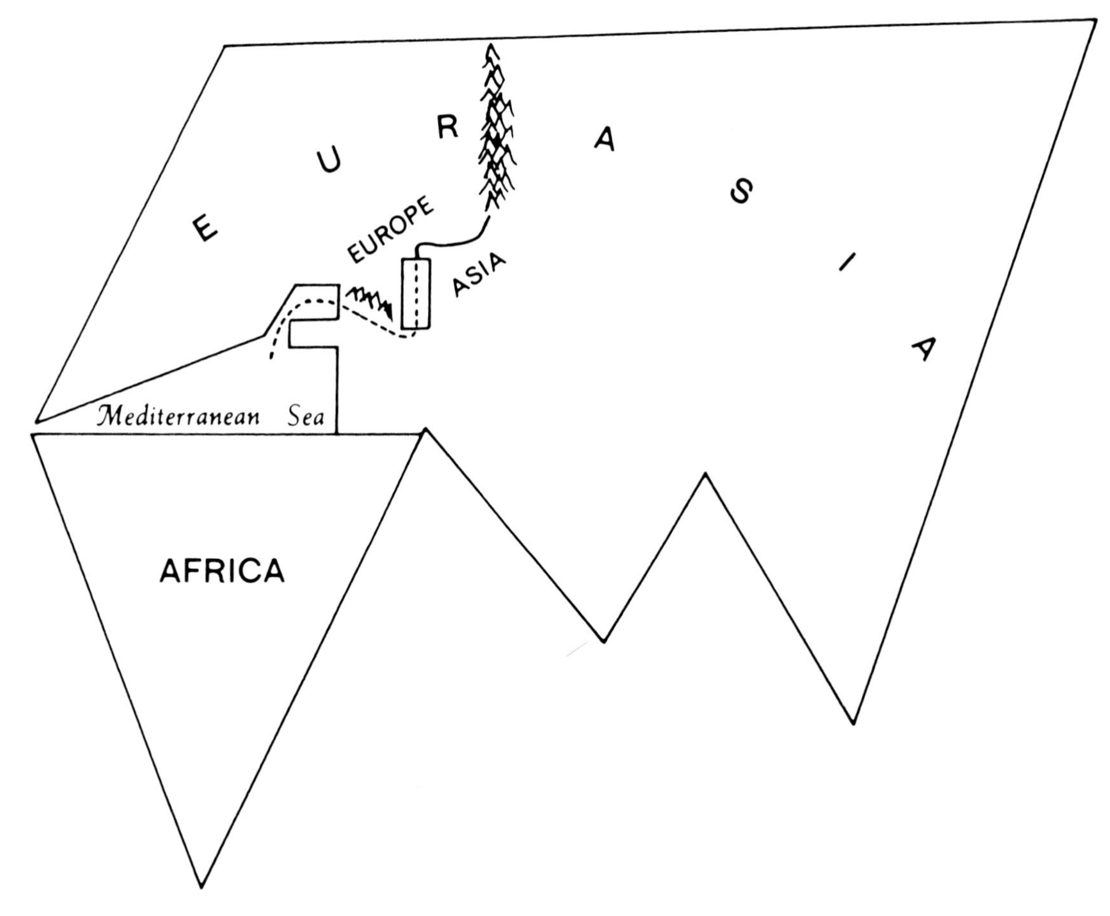

Self-Test

1. What is the name of the north-south mountain range that divides Eurasia into Europe and Asia?

   ____

2. What is the name of the east-west mountain range that helps define part of the Europe-Asia boundary?

   ____

3. What *two* seas are *half* in Europe and *half* in Asia?

   ______________________ and ______________________

4. What sea separates Europe from Africa? ______________________

Check page 5 for the correct answers before continuing.

This completes a general survey of the oceans, continents, a few seas, mountains, and a small number of terms. At this point, you should be able to go to a map and quickly and easily locate any of the features mentioned in the preceding pages. It would be a good idea to review this first chapter and once again study the world map before moving on. If this material is firmly set in mind at this stage, GO ON.

2

North America

North America, one of the six continents, may be divided for study into *Canada and the United States,* and *Middle America.* The Canadian–United States portion of North America is sometimes called *Anglo-America.* A linguistic definition such as this (Anglo = English) satisfactorily denotes the majority pattern, but obscures many areas where English is not the majority language. *Greenland,* a detached piece of the North American Continent, is discussed at the end of part I, section A.

I. Canada and the United States

A. CANADA

For the citizen of the United States, Canada is the giant neighbor to the north (unless one lives in Alaska or Hawaii; Canada is then to the east or northeast). In *area,* Canada is larger than the United States (about the size of the 50 states plus one more the size of Texas) and second only to the USSR among the world's countries. In *population,* the United States is almost ten times as large as Canada.

Country	*Area*		*Population*
	km²	*mi²*	*(1988)*
USSR	22,272,200	8,599,300	285,000,000
Canada	9,972,334	3,851,809	26,000,000
United States	9,519,622	3,675,547	245,000,000

For administrative purposes, Canada is divided into *two territories* and *ten provinces.* The two territories are sparsely populated, far-northern areas, called *Yukon Territory* and *Northwest Territories.* Their names suggest their locations. Look at the atlas map of Canada and find the *Yukon* and *Northwest Territories* and fix their locations in your mind.

Name the two Canadian territories:

1. ____________________ and 2. ____________________

There are *ten* provinces in Canada. From the Pacific Ocean to the Atlantic Ocean, the *seven* large ones are:

1. BRITISH COLUMBIA
2. ALBERTA
3. SASKATCHEWAN
4. MANITOBA
5. ONTARIO
6. QUEBEC
7. NEWFOUNDLAND (includes Labrador)

The *Maritime Provinces* of Canada lie northeast of the United States, and one of them, New Brunswick, shares a common boundary with Maine.
When the study of the atlas and the map below is completed, take the self-test that follows.

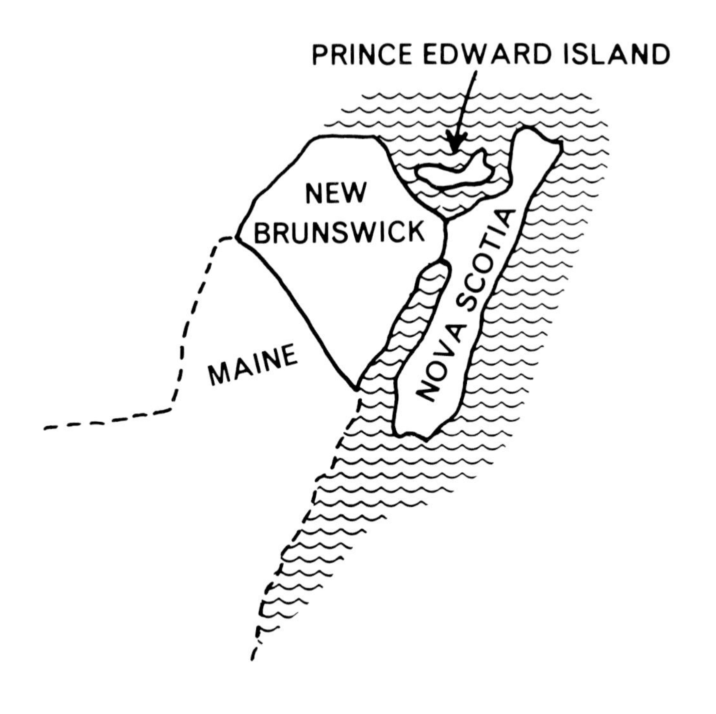

Self-Test

The two Canadian territories are ______________________________

____

The seven *large* Canadian provinces are ______________________________

____

____

____

____

The three *small* Canadian provinces are ______________________________

____

____

Check the answers with pages 7–8 before continuing.

There are some key Canadian cities that everyone should know. These are given below and the capital cities are in italics. Open the atlas and locate the capitals while proceeding through the list.

Territories:

1. Yukon Territory—*Whitehorse*
2. Northwest Territories—*Yellowknife*

The Four Large Provinces of Western Canada:

1. British Columbia—*Victoria,* Vancouver

   Avoid an error here that almost everyone makes! *Victoria* is located on Vancouver Island; Vancouver is NOT located on the island of the same name!

   When thinking of British Columbia, think *first* of Vancouver. This beautiful city is the largest in the province and may be reached overland from the United States by rail or highway.

   Vancouver Island is a large island (over 480 km [300 mi] long), lying offshore from the city of Vancouver. The provincial capital, *Victoria,* is on Vancouver Island.

2. Alberta—*Edmonton,* Calgary

   Cal is in *Al* (i.e., *Cal*gary is in *Al*berta). Is there some way to remember *Edmonton,* the province's capital and largest city?

3. Saskatchewan—*Regina,* Saskatoon

   The Indian name for this province is easy to spell because it is absolutely phonetic: Sas-katch-e-wan. *Sas*(katoon) just seems to belong in *Sas*(katchewan).

   Here again, the capital is the largest city and the more difficult city to remember. Fix the name *Regina* in mind.

4. Manitoba—*Winnipeg*

   This city of the Canadian prairie is the largest in central Canada. When thinking of *Mani*(toba), think of *Winni*(peg).

Now, review the western half of Canada by filling in the blanks.

Territory	*Capital*	*Major City*
Yukon	____________________	omit
Northwest Territories	____________________	omit

Province		
British Columbia	____________________	__________
Alberta	____________________	__________
Saskatchewan	____________________	__________
Manitoba	____________________	omit

Cover the answers you have written and turn things around and try once more before looking up the correct answers.

Capital	*Major City*	*Province or Territory*
Whitehorse	omit	________________
Yellowknife	omit	________________
Victoria	Vancouver	________________
Edmonton	Calgary	________________
Regina	Saskatoon	________________
Winnipeg	omit	________________

NOW, check your answers above in both sections.

Minimum performance requires that *all* of the provinces, territories, capitals, and major cities to this point are learned. Correct any mistakes *before* continuing.

You have now mastered the two Canadian territories and the four large western provinces and their cities. *Three large eastern provinces,* the *three maritime provinces,* and their cities remain to be learned.

The provinces and their important cities are:

1. Ontario—*Toronto,* Windsor, London, Hamilton, *OTTAWA.*

   The first four of these cities are located on that southern prong of Canada that jabs through the eastern Great Lakes.

   OTTAWA is the Canadian *national* capital, located in Ontario across the Ottawa River from Québec. This is the linguistic boundary between English-speaking Ontario and French-speaking Québec. The National Capital Region, however, encompasses Hull, Québec, and its immediate vicinity. Of this National Capital Region, 2,720 km^2 (1,050 mi^2) are in Ontario; 1,943 km^2 (750 mi^2), in Québec. In many aspects, Ottawa-Hull functions as a federal district and from a geographical viewpoint is excellently situated.

2. Québec—*Québec,* Montréal, Hull

   The capital of Québec province is easy to remember—*Québec.* Montréal is the largest city in Canada (though the Toronto *metropolitan area* is larger).

3. Newfoundland—*St. John's*

   Newfoundland is the only Canadian province that consists of *two* distinct landmasses: (1) *the island of Newfoundland,* and (2) *mainland Labrador.*

   It seems appropriate that a two-part province have a capital with *two words*—St. John's.

   Newfoundland, including Labrador, was probably the first area in North America visited by Europeans (by the Vikings in the tenth century). Its name is most fitting.

4. New Brunswick—*Fredericton,* Saint John

   There is no gimmick for remembering *Fredericton* as New Brunswick's capital. To add to the problem, Saint John, the largest city of the province, is easily confused with *St. John's,* the capital of Newfoundland.

5. Nova Scotia—*Halifax*

   Again, the name must be memorized.

6. Prince Edwards Island—*Charlottetown*

   The smallest province in Canada has the longest name for its capital city.

Now, review the eastern half of Canada.

Province	*Capital*	*Major City*
Ontario	____________	____________
		____________
		____________
		____________ (national capital)
Québec	____________	____________
		____________
Newfoundland	____________	omit
New Brunswick	____________	____________
Nova Scotia	____________	omit
Prince Edward Island	____________	omit

Now, cover your answers above and reverse the procedure.

Capital	*Major City*	*Province*
Toronto	Windsor	
	London	
	Hamilton	
	OTTAWA	____________
Québec	Montréal	
	Hull	____________
St. John's	omit	____________
Fredericton	St. John	____________
Halifax	omit	____________
Charlottetown	omit	____________

Return to pages 9–10 and check these answers. *Minimum* performance requires knowing *all* of the provinces, capitals, and major cities. Correct any errors before turning the page.

In addition to the territories, provinces, capitals, and major cities of Canada, there are certain other localities that should be noted. The following should be located in the atlas, one by one, while going through the list. Pause with each one and fix its location firmly in mind before going to the next.

Salt Water Bodies:

Hudson Bay

Associated with Hudson Bay are:

James Bay (southern extension of Hudson Bay)
Foxe Basin (northern extension of Hudson Bay)
Hudson Strait (eastern exit of Hudson Bay)
Baffin Island (north of Hudson Bay, Foxe Basin, Hudson Strait)
Baffin Bay (between Baffin Island and Greenland)
Davis Strait (between Baffin Bay and the North Atlantic)

Gulf of St. Lawrence (between Newfoundland and New Brunswick)
Bay of Fundy (between southern New Brunswick and Nova Scotia; famous for tides up to 15 m [50 ft.])

Lakes:

The Great Lakes	*U.S. Area*		*Canadian Area*	
	km²	*mi²*	*km²*	*mi²*
1. Superior	53,820	20,700	28,860	11,100
2. Huron	23,660	9,100	36,140	13,900
3. Erie	12,948	4,980	12,818	4,930
4. Ontario	9,360	3,600	10,400	4,000
5. Michigan	58,240	22,400	(none; entirely in U.S.)	

Lake Winnipeg (north of Winnipeg, in Manitoba)

Great Slave Lake]
Great Bear Lake] —Northwest Territories

Rivers:

St. Lawrence River (from Lake Ontario to the Gulf of St. Lawrence)

Fraser River (British Columbia; Vancouver is at its mouth)

Mackenzie River (flows north into the Arctic Ocean through the Northwest Territories)

Open the atlas to the best map of Canada it contains and settle back in a comfortable chair. See how many of the items studied can be readily located on the map. Look at other features, especially those not covered in this section, such as mountains, plains, and islands. Turn to other maps of Canada, North America, and the World, and look at the climate, vegetation, economy, population, and similar maps. Review these maps carefully and leisurely.

Do not attempt to go on to the section on the United States until confident that this information about Canada is thoroughly learned.

If you think you know Canada, label the map on the following page, using ALL the names you have learned to this point.

TWO INTERESTING NEIGHBORS OF CANADA

1. St. Pierre and Miquelon. These two islands of 6,000 people and 241 rocky square km (93 mi²) are relics of France's once vast holdings in North America. They lie about 19 km (12 mi) south of Newfoundland and comprise an *Overseas Department* of France. In other words, the department is an integral part of the French state.
2. Greenland. The world's largest island (unless you call Australia an island) is culturally Eskimo, politically Danish, and physically a detached piece of North America. A mere 65 million years ago it was still joined to the continent. Greenland (*Kalaallit Nunaat* in Eskimo) is a self-governing part of the Kingdom of Denmark. *Nûk* (formerly Godthåb) is the capital. Eskimo place-names became official in 1980.

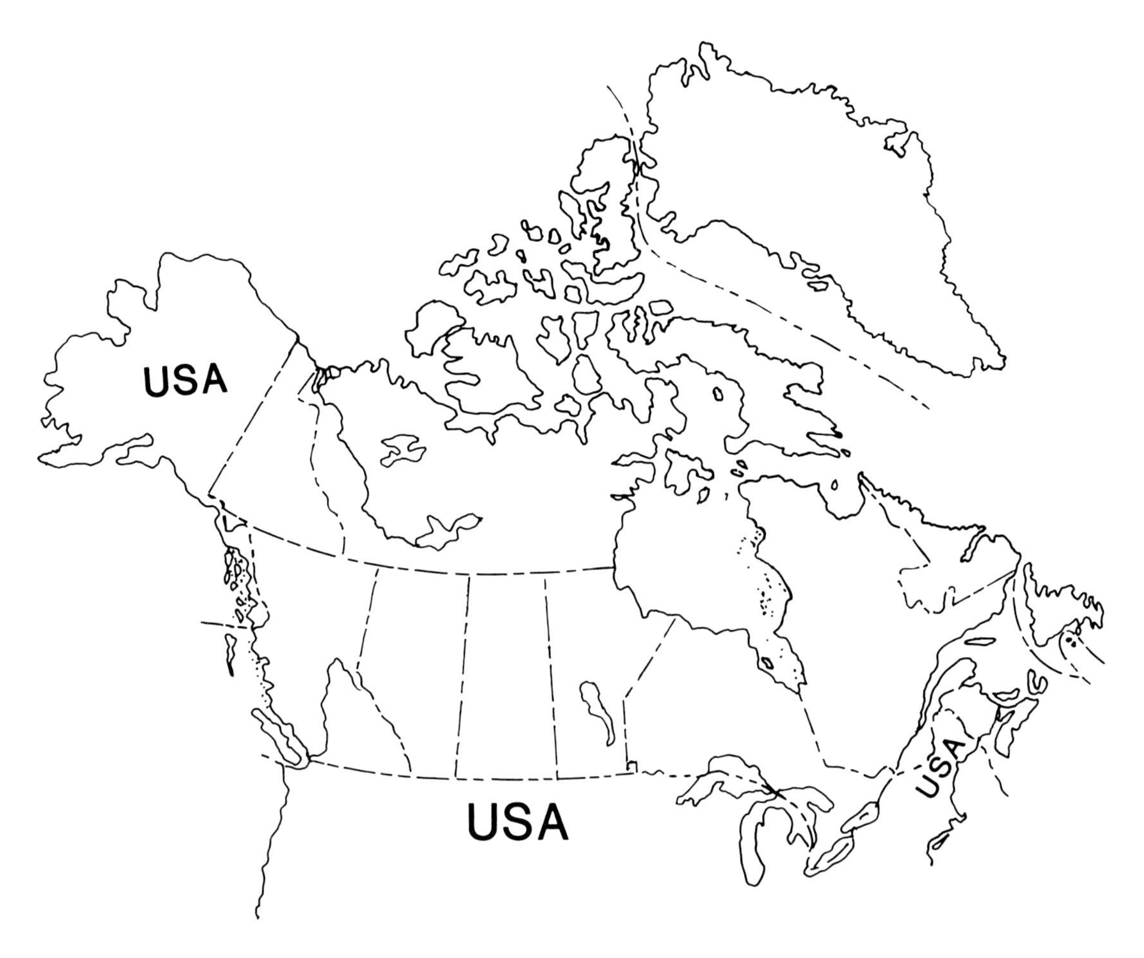

Study the map below, then quickly and accurately locate the named features in the atlas.

Iceland is normally considered to be the westernmost part of Europe, but it is shown below for comparative purposes. *Baffin Island* is part of Canada's Northwest Territories.

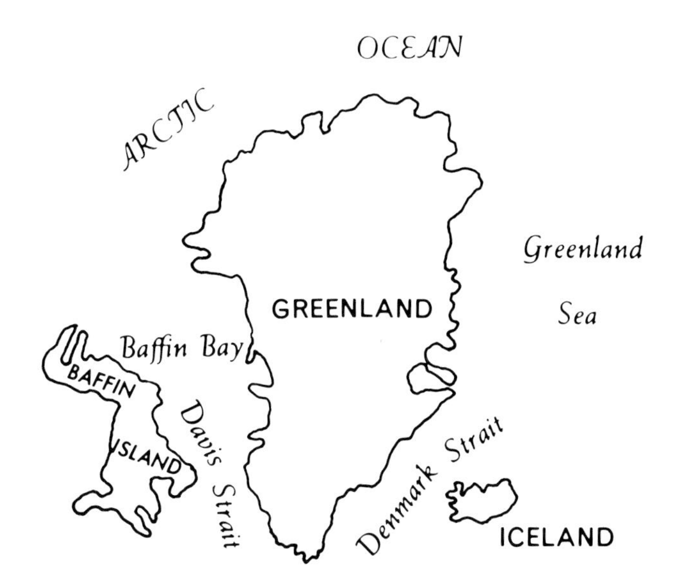

B. UNITED STATES

The United States is part of North America and, with Canada, occupies most of that continent. Everyone is expected to learn the 50 states of the Union and the District of Columbia. These should be located on the map—quickly and easily. Everyone should also be able to list them without the aid of a map.

The Seven Major Sections of the United States:

1. New England (6 states)
2. Middle Atlantic (5 states and the District of Columbia)
3. South (11 states of the old Confederacy)
4. Middle West (10 states)
5. Great Plains (5 states)
6. Rocky Mountain (8 states)
7. Pacific (5 states)

These seven areas are located on the cartogram below.

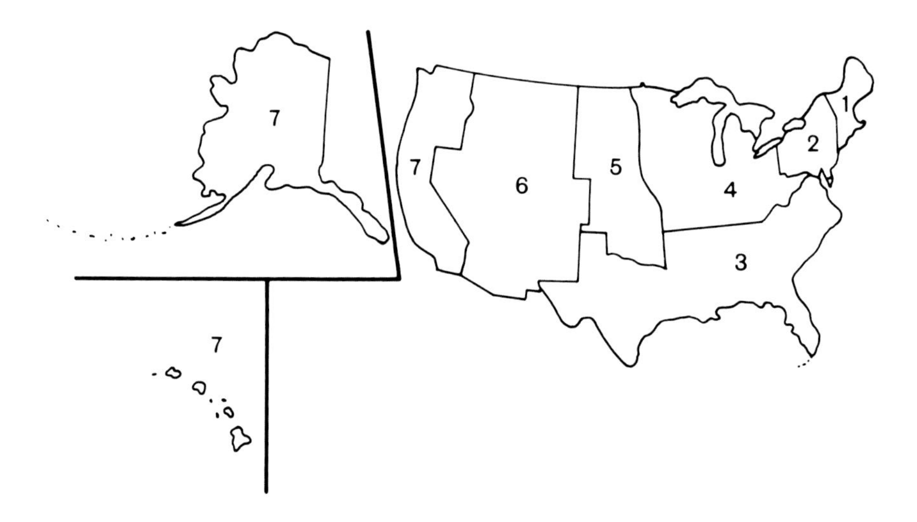

Self-Test

The two main components of "Anglo-America" are ______________________

______________________ and ______________________.

The three main components of North America are ______________________

______________________, ______________________,

and ______________________.

The seven major areas of the United States are ______________________________

____

____

____

____

____

Return to pages 7 and 14 in order to check these answers. *Make certain* the answers above are correct before continuing.

The Seven Major Sections of the United States:

1. *NEW ENGLAND* (the North East)
   Maine
   New Hampshire (Many can't remember if this is coastal or inland; it contains a city named *Portsmouth,* therefore *must* be coastal.)
   Vermont
   Massachusetts (Learn to spell this one! It is perfectly phonetic—*Mass-a-chu-setts.*)
   Connecticut (This is also easy to spell, but be sure to use only *one* "t" at the end! *Connect-i-cut.*)
   Rhode Island (This is the smallest U.S. state, and NOT an island.)

Review this list, find the places in the atlas, and ponder over them a little before going to the second group.

---

2. *MIDDLE ATLANTIC* (*midway* between New England and the South)
   New York
   New Jersey
   Pennsylvania
   Delaware
   Maryland
   District of Columbia (Washington, D.C.)

DON'T HURRY! Look over *both* lists above before continuing.

3. *SOUTH* (The 11 states of the old Confederate States of America)
   Virginia
   North Carolina
   South Carolina
   Georgia
   Florida
   Alabama
   Mississippi
   Louisiana (Pronounce this correctly! Lou-*eez*-ee-ana.)
   Texas
   Arkansas]—the only inland states of this area
   Tennessee]

As before, look at these on the map before plunging ahead!

---

4. *MIDDLE WEST* (This may be a good historical name but today it is a poor geographical term; tradition, however, compels its usage.)
   West Virginia
   Kentucky
   Missouri
   Ohio
   Indiana
   Illinois
   Iowa
   Michigan (Don't overlook that part of Michigan attached to Wisconsin.)
   Wisconsin
   Minnesota
5. *GREAT PLAINS* (This is the wheat and cattle country of mid-United States.)
   North Dakota
   South Dakota
   Nebraska
   Kansas
   Oklahoma
6. *ROCKY MOUNTAIN* (Most people call this area the "West.")
   Montana
   Idaho
   Wyoming
   Colorado
   Utah
   Nevada
   New Mexico
   Arizona
7. *PACIFIC* (These states all border the Pacific Ocean.)
   California
   Oregon
   Washington
   Alaska
   Hawaii

Review the seven sections and their states. Take a last look at them in the atlas. When ready, take the self-test below.

Self-Test

Name all of the states by sections.

1. *New England:* ______________, ______________, ______________, ______________,
   ______________, ______________

2. *Middle Atlantic:* ________________, ________________, ________________, ____________,

    ________________ and the nation's capital ________________

3. *South:* ________________, ________________, ________________, ________________,

    ________________, ________________, ________________, ________________,

    ________________, ________________, ________________

4. *Middle West:* ________________, ________________, ________________, ________________,

    ________________, ________________, ________________, ________________,

    ________________, ________________

5. *Great Plains:* ________________, ________________, ________________, ________________,

    ____

6. *Rocky Mountain:* ________________, ________________, ________________, ____________,

    ________________, ________________, ________________, ________________

7. *Pacific:* ________________, ________________, ________________, ________________,

    ____

Check all of the answers by turning to pages 15 and 16. Correct *all* errors. *DO NOT* continue until certain of all the states!

When ready, label the 50 states on the map provided on page 19.

There is at least one city in every state that should be known (the capital), and there are other important cities that it would be worthwhile to know. Below are listed the capitals (in italics) and a *few* of the important cities. Many major United States cities are omitted, and many small but important towns are omitted. No attempt is made to justify this list.

1. *NEW ENGLAND*
    - Maine .... *Augusta,* Portland
    - New Hampshire .... *Concord,* Portsmouth, Manchester
    - Vermont .... *Montpelier,* Burlington
    - Massachusetts .... *Boston,* Worcester, Springfield
    - Connecticut .... *Hartford,* New Haven
    - Rhode Island .... *Providence*
2. *MIDDLE ATLANTIC*
    - New York .... *Albany,* New York, Buffalo, Rochester, Syracuse
    - New Jersey .... *Trenton,* Newark, Jersey City
    - Pennsylvania .... *Harrisburg,* Philadelphia, Pittsburgh
    - Delaware .... *Dover,* Wilmington
    - Maryland .... *Annapolis,* Baltimore
    - District of Columbia .... *Washington*

3. *SOUTH*

Virginia .... *Richmond,* Norfolk
North Carolina .... *Raleigh,* Charlotte, Greensboro, Winston-Salem
South Carolina .... *Columbia,* Charleston
Georgia .... *Atlanta,* Savannah
Florida .... *Tallahassee,* Miami, Tampa, Jacksonville
Alabama .... *Montgomery,* Birmingham, Mobile
Mississippi .... *Jackson*
Louisiana .... *Baton Rouge,* New Orleans
Texas .... *Austin,* Houston, Dallas, San Antonio, El Paso
Arkansas .... *Little Rock*
Tennessee .... *Nashville,* Memphis

4. *MIDDLE WEST*

West Virginia .... *Charleston*
Kentucky .... *Frankfort,* Louisville
Missouri .... *Jefferson City,* St. Louis, Kansas City
Ohio .... *Columbus,* Cleveland, Cincinnati, Toledo, Akron
Indiana .... *Indianapolis,* Gary, South Bend, Fort Wayne
Illinois .... *Springfield,* Chicago
Iowa .... *Des Moines*
Michigan .... *Lansing,* Detroit
Wisconsin .... *Madison,* Milwaukee
Minnesota .... *St. Paul,* Minneapolis, Duluth

5. *GREAT PLAINS*

North Dakota .... *Bismarck,* Fargo
South Dakota .... *Pierre,* Sioux Falls
Nebraska .... *Lincoln,* Omaha
Kansas .... *Topeka,* Kansas City, Wichita
Oklahoma .... *Oklahoma City,* Tulsa

6. *ROCKY MOUNTAIN*

Montana .... *Helena,* Great Falls
Idaho .... *Boise,* Pocatello
Wyoming .... *Cheyenne*
Colorado .... *Denver,* Boulder, Pueblo
Utah .... *Salt Lake City,* Ogden, Provo
Nevada .... *Carson City,* Reno, Las Vegas
New Mexico .... *Santa Fe,* Albuquerque
Arizona .... *Phoenix,* Tucson

7. *PACIFIC*

California .... *Sacramento,* Los Angeles, San Francisco, San Diego
Oregon .... *Salem,* Portland
Washington .... *Olympia,* Seattle, Spokane
Alaska .... *Juneau,* Anchorage, Fairbanks
Hawaii .... *Honolulu*

It is *not* essential that every city listed be known. Quite a few on the list, and perhaps others not listed, are familiar. It is impossible to cite every city in the United States, and who is to say what city is significant? Review these, however, and look them up in the atlas. A personal inventory of familiar place-names will expand with time.

WATER BODIES

The largest lakes in the United States (or partially within the United States) are *THE GREAT LAKES* (see page 12).

Other large United States lakes are:

Great Salt Lake, Utah	(salt)	3,900 km² (1,500 mi²)
Lake Iliamna, Alaska	(fresh)	2,750 km² (1,057 mi²)
Lake Okeechobee, Florida	(fresh)	1,900 km² (730 mi²)
Lake Pontchartrain, Louisiana	(salt)	1,625 km² (625 mi²)

Each student should learn these lakes by name and location.

Seas, Gulfs, Bays, Sounds, and Straits:

Four important water bodies are adjacent to the 48 states.

1. Gulf of Mexico (southeast of the mainland)
2. Straits of Florida (between Florida and Cuba/Bahamas)
3. Chesapeake Bay (Maryland/Virginia)
4. Puget Sound (Washington)

Four additional water bodies are associated with Alaska.

1. Gulf of Alaska (south of the Alaskan mainland)
2. Bering Sea (southwest of Alaska proper, north of Aleutians)
3. Bering Strait (between Alaska and USSR)
4. Beaufort Sea (north of Alaska, Yukon, and Northwest Territories)

Rivers:

Mississippi River ⎤
Missouri River ⎥ .... midsection of United States
Ohio River ⎦
Arkansas River
Hudson River .... New York
Colorado River .... southwest United States
Rio Grande
Columbia River ⎤ .... Washington/Oregon
Yukon River ⎦ .... Alaska/Yukon Territory

Before continuing, review *all* of the water bodies.

HIGHLAND AREAS

Each one listed should be readily located.

1. *New England Highlands*—northern New York, New England
2. *Appalachian/Blue Ridge*—northern Alabama to southern New York
3. *Ozarks/Ouachita Mountains*—Arkansas, Oklahoma, Missouri
4. *Rocky Mountains*—New Mexico through Idaho/Montana to Yukon
5. *Sierra Nevada/Cascade Mountains*—eastern California (Sierra Nevada); central Oregon and Washington (Cascade Mountains)
6. *Coast Ranges*—coastal California, Oregon, Washington
7. *Brooks Range*—northern Alaska
   The *North Slope* oil fields lie on the north slope of the Brooks Range, centering on the town of Prudhoe Bay.
8. *Alaska Range*—southern Alaska (contains Mt. McKinley, highest mountain in the United States—6,193 m [20,320 ft.])

With the help of an atlas, label the above features and those mentioned on page 19. Use the map on page 21. Certain features, such as rivers and mountains will have to be drawn on the map.

II. Middle America

Middle America is part of *North America,* and it is also part of *Latin America.* Middle America is *NOT* part of *South America.*

Refer to the cartogram on page 22.

1. Label *Middle America.*
2. Label the oceans on either side of the continents.
3. Label the *Caribbean Sea* (a part of Middle America).

Study carefully, then take the test on the following page.

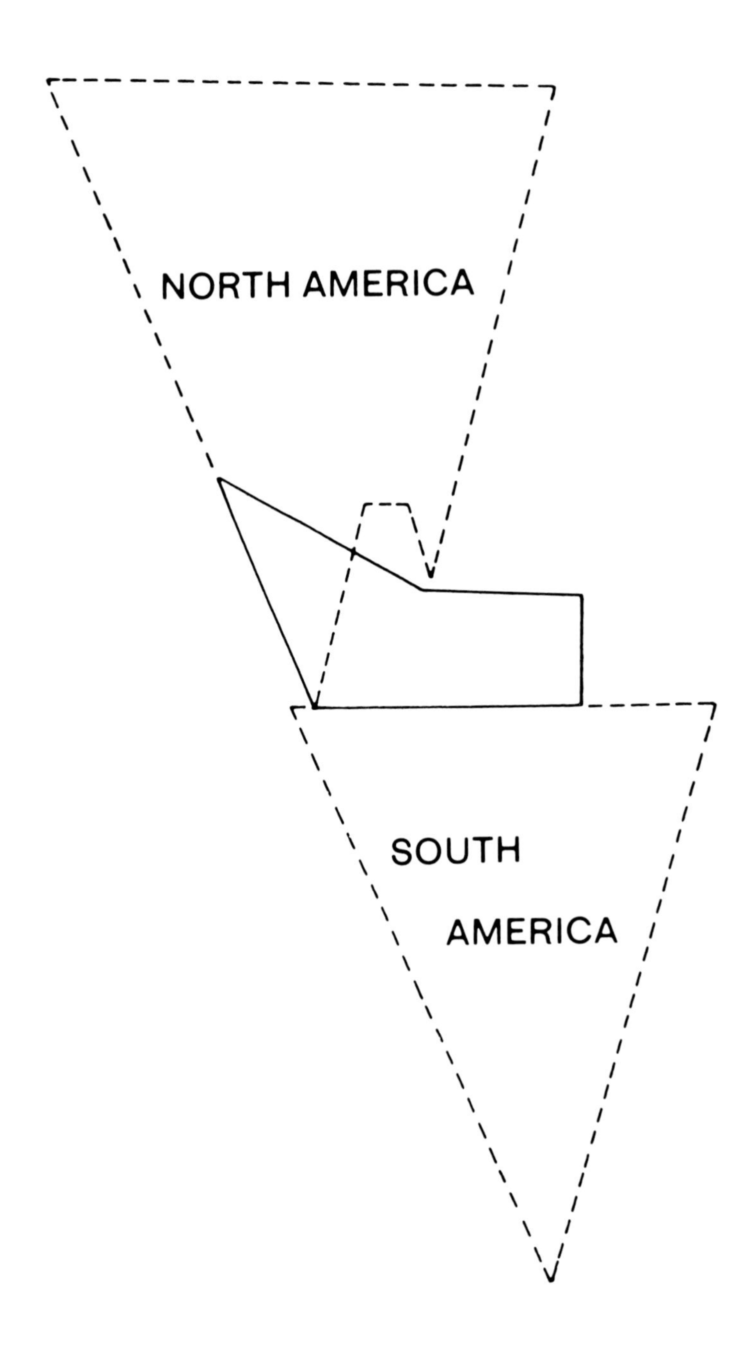

Self-Test

(Check true or false)

	True	False
1. Middle America *includes* part of South America.	_____	_____
2. Middle America extends into the Atlantic Ocean.	_____	_____
3. The Caribbean Sea is an extension of the Atlantic Ocean.	_____	_____

4. Middle America is a continent. ______ ______

5. Middle America is a part of North America. ______ ______

6. Middle America is a part of South America. ______ ______

Answers for above: (1) F, (2) T, (3) T, (4) F, (5) T, (6) F. If one or more are missed, return to page 21 and continue from there.

Study the cartogram at the right. *Central America* is that part of Middle America that *excludes* Mexico and the West Indies (Antilles).

1. Label Mexico.
2. Label the West Indies.
3. Label Central America.
4. Label the Caribbean Sea.

Turn to page 24 for answers.

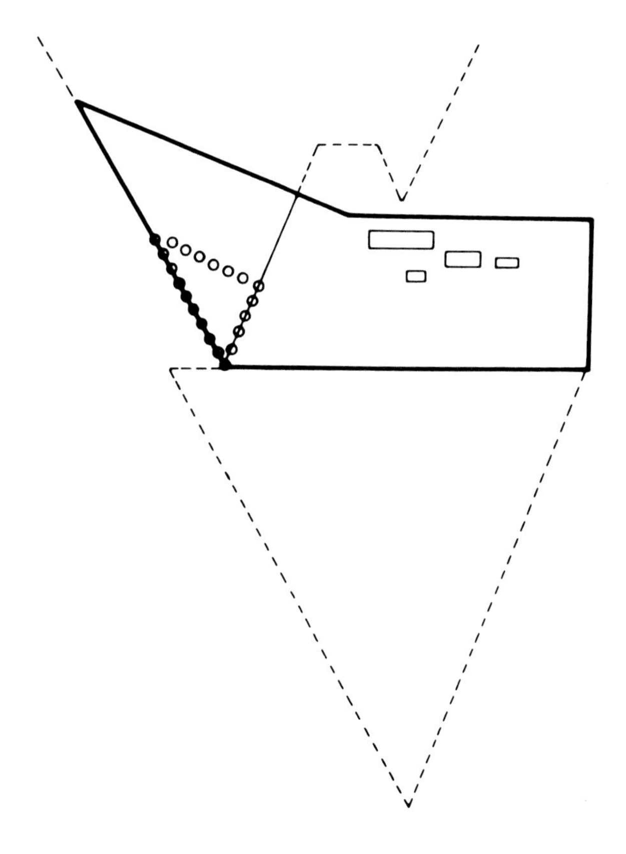

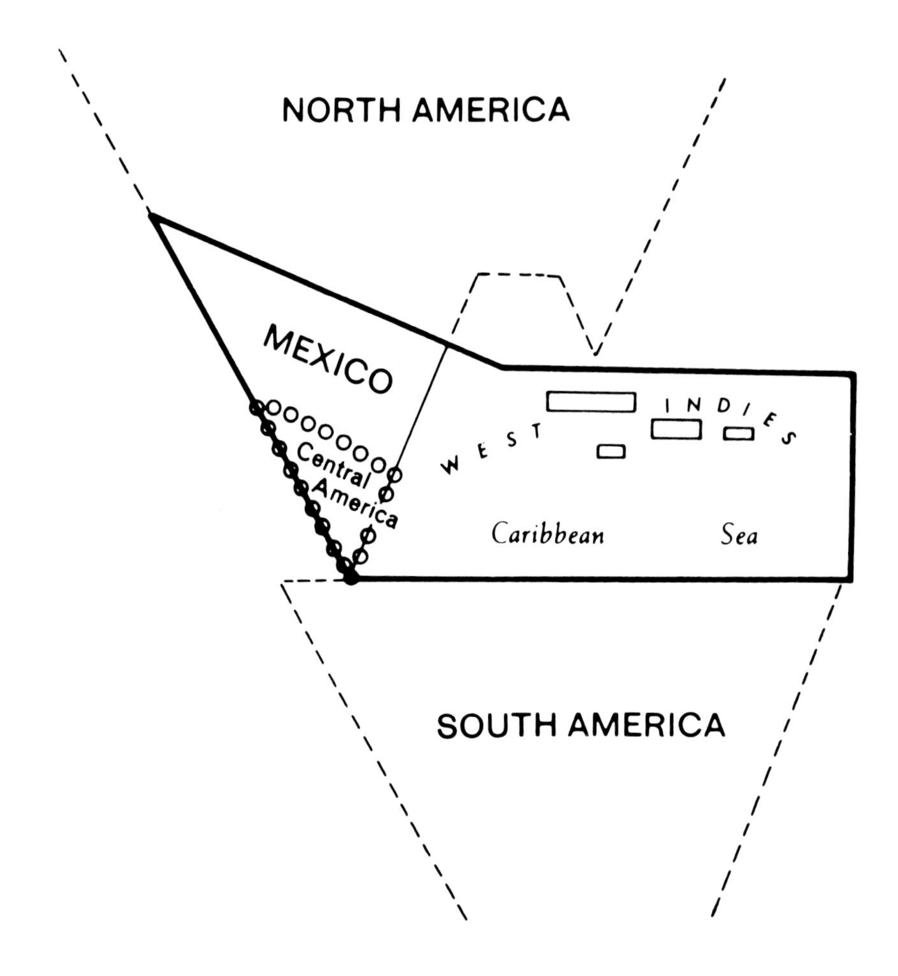

With the atlas closed, label the items below on the next page:

1. North America
2. South America
3. Middle America
4. Central America
5. West Indies
6. Pacific Ocean
7. Arctic Ocean
8. Atlantic Ocean
9. Caribbean Sea
10. Mexico

Open the atlas to a map of the Caribbean area. Name each of the four large islands of the West Indies (label on p. 25). Check this work with the atlas, then continue.

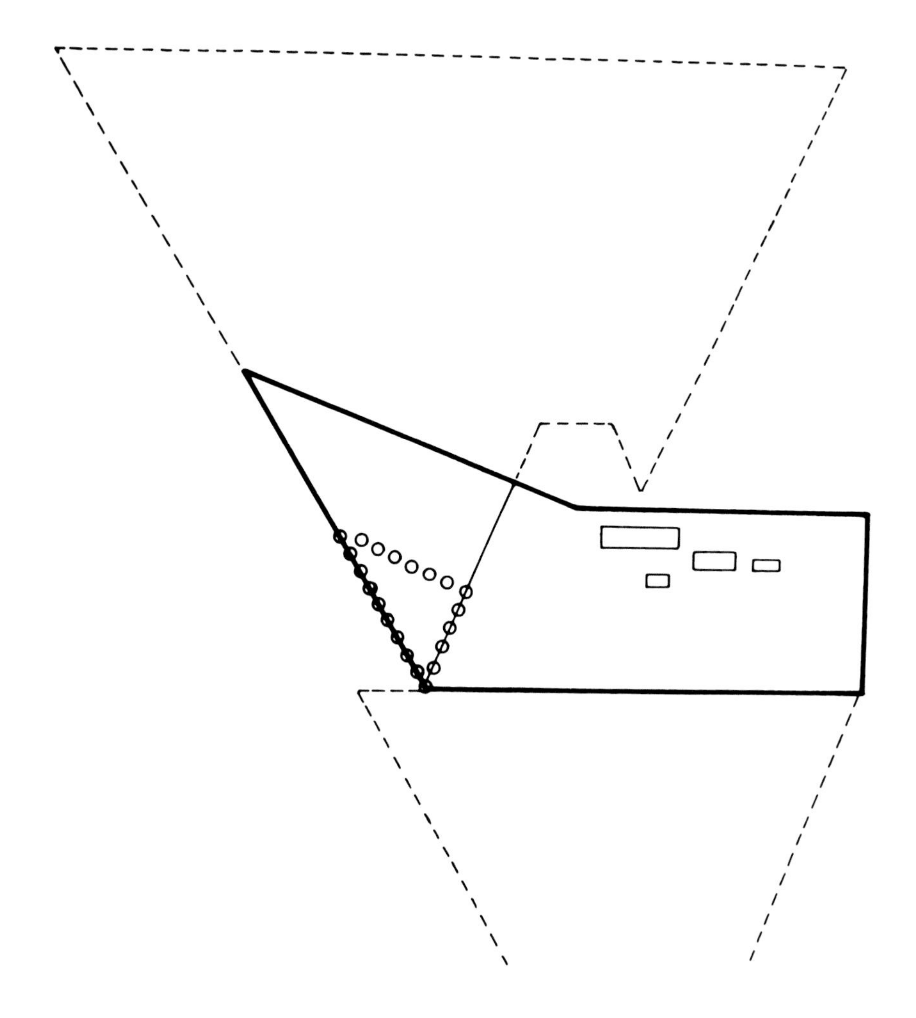

Are the four large islands correctly labeled above? The largest is *Cuba;* directly south is *Jamaica;* east of those two is *Hispaniola* (*NOT* Haiti or Dominican Republic—these are *countries* on the island of Hispaniola); east of Hispaniola is *Puerto Rico.*

Self-Test

(Check true or false)

	True	*False*
1. Mexico is part of North America.	_____	_____
2. Mexico is part of Middle America.	_____	_____
3. Mexico is part of Central America.	_____	_____
4. Mexico is part of South America.	_____	_____

5. The West Indian islands are part of Middle America. ______ ______

6. The West Indian islands are part of Central America. ______ ______

7. Central America is part of North America. ______ ______

8. Central America is part of Middle America. ______ ______

9. Central America is part of South America. ______ ______

Answers for above: (1) T, (2) T, (3) F, (4) F, (5) T, (6) F, (7) T, (8) T, (9) F. If *all* of the preceding are answered correctly, continue the study. If *even one* is missed, return to pages 24 and 25 and REVIEW before moving on.

Middle America is commonly divided into *three* distinct areas:

A. Mexico
B. Central America
C. Caribbean (West Indies)

Study the cartogram on the next page carefully before proceeding.

Self-Test

Middle America is commonly divided into the following areas:

A. ______________________________

B. ______________________________

C. ______________________________

Check the answers above and correct any errors before continuing.

A. MEXICO

Mexico, southern neighbor of the United States, is a large country composed of 31 *states* and a *Federal District.* Its official name is *Estados Unidos Mexicanos* (United Mexican States).

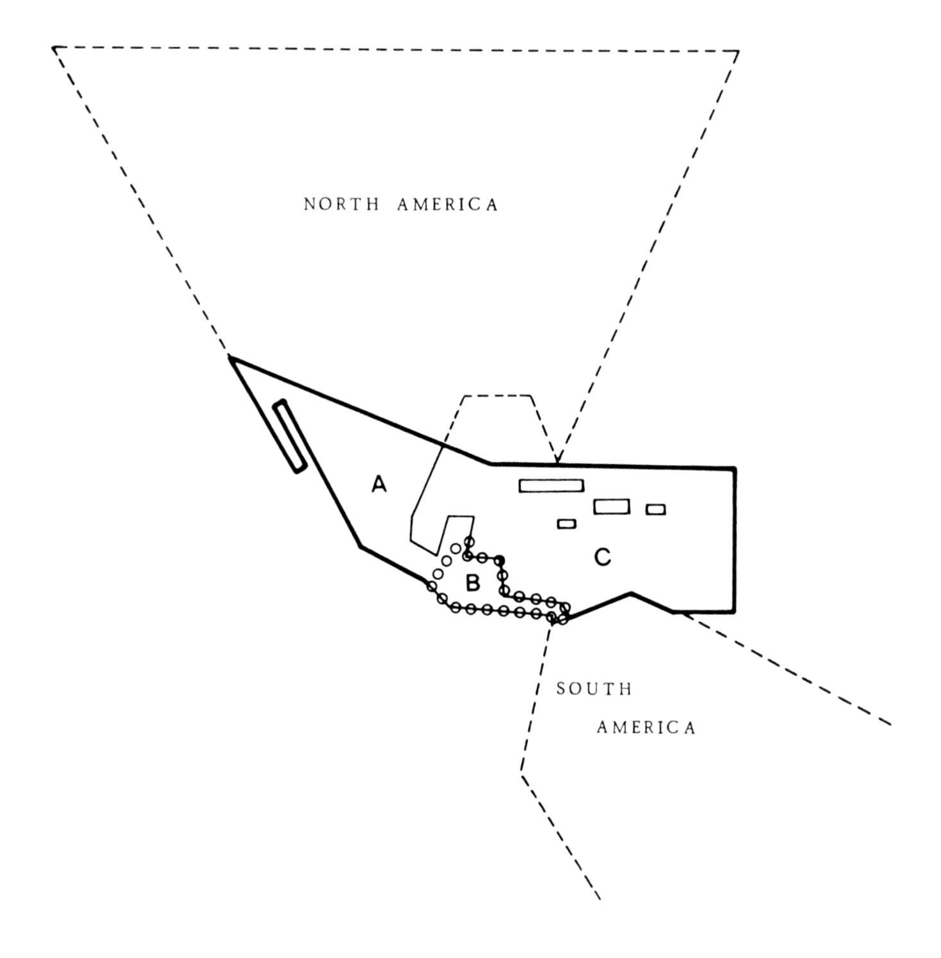

A *minimum* knowledge of Mexican locations requires knowing the following:

Cities: *México (Mexico City)*
Monterrey
Guadalajara
Mérida
Veracruz
Tijuana
Ciudad Juárez
Nuevo Laredo
Chihuahua
Acapulco

Peninsulas: Yucatán
Baja California

Isthmus: Tehuantepec

Mountains: Sierra Madre Occidental
Sierra Madre Oriental
Sierra Madre del Sur

River: Río Grande (more properly, Río Grande del Norte, but called Río Bravo in Mexico)

Salt Water Bodies: Pacific Ocean
Gulf of California
Gulf of Mexico
Bay of Campeche

Neighbors: United States
Guatemala
Belize

Locate *each* of the above (and those mentioned on page 27) on the cartogram. Use the atlas for assistance.

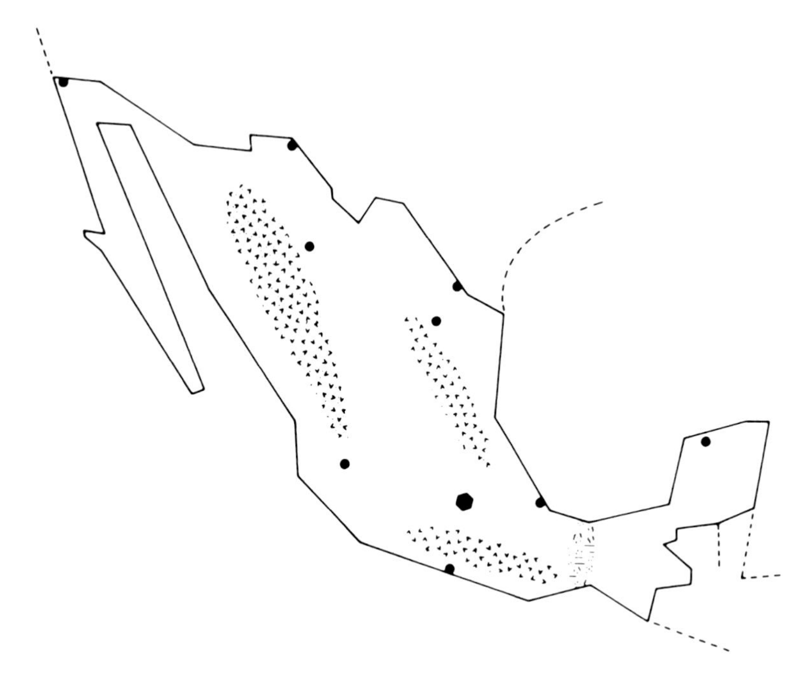

Self-Test

The *three* distinct areas into which Middle America is normally divided are

A. ______________________________

B. ______________________________

C. ______________________________

The capital city of Mexico is ______________________________

Nine other important Mexican cities are __________, __________, __________,

__________, __________, __________, __________, __________,

____

Two Mexican peninsulas are __________ and __________

An isthmus in Mexico is ______________________________

Three Mexican mountain ranges are ______________________________,

__________________, __________________

An important Mexican river is ____________________

Three gulfs/bays adjacent to Mexico are ____________________,

____________________ and ____________________

Check the answers on pages 26–27 *before* continuing.

B. CENTRAL AMERICA

Central America is part of *Middle America* and part of *North America*. It includes *seven* independent countries.

1. Guatemala
2. Belize
3. El Salvador
4. Honduras
5. Nicaragua
6. Costa Rica
7. Panama

Locate these in the atlas and name *each* on the cartogram below. Label the two major water bodies associated with Central America. Name the two lakes shown on the cartogram. Locate the Panama Canal.

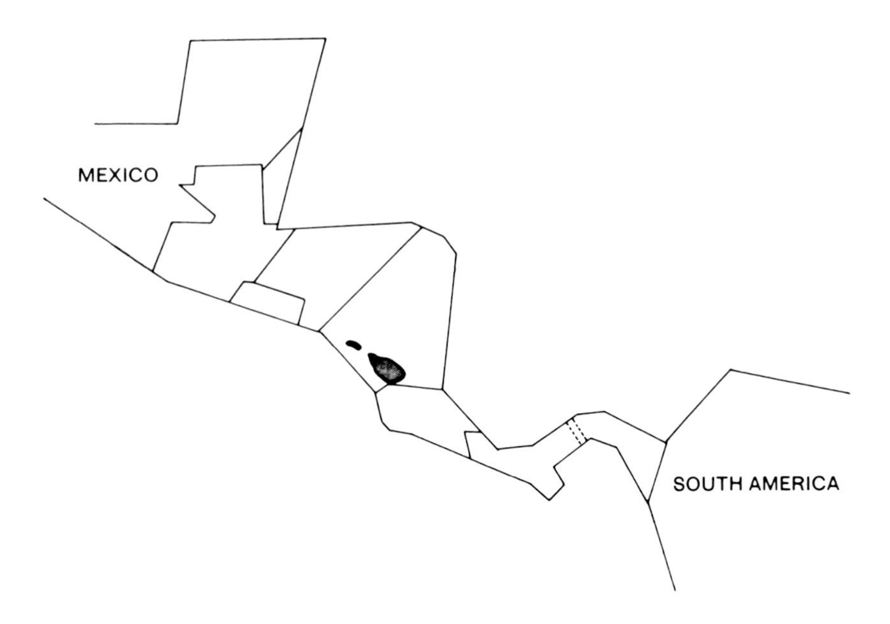

When *sure* of the names and locations of the seven political units and the four water bodies, go on to the self-test.

Self-Test

Name the seven political units of Central America.

__________, __________, __________, __________,

__________, __________, __________.

Name the water body that lies northeast of Central America. ________________ .

Name the water body that lies southwest of Central America. ________________ .

Name the largest lake in Central America. ________________________

This lake is in the country of ____ .
Check your answers with page 29. Make sure the countries and water bodies are listed correctly before continuing.

Refer to the atlas and locate each capital city. List each country and its capital below.

Country	*Capital City*
________________________	________________________
________________________	________________________
________________________	________________________
________________________	________________________
________________________	________________________
________________________	________________________
________________________	________________________

Study the above list carefully. Note that in *two* instances the capital is named for the country, and in another instance a country and its capital have *part* of the name in common.

Self-Test

Name the three main divisions of Middle America.

A. ____

B. ____

C. ____
Check these answers and make certain they are correct before continuing.

C. CARIBBEAN (WEST INDIES)

Compare the outline map on page 32 with an atlas map of this area and complete the following blanks.

1. *Four* large islands comprise the *Greater Antilles.* They are ________________ , ____________ , ________________ and ________________ .

2. The large peninsula north of Cuba is ________________ .

3. The group of islands between the Greater Antilles and Florida is called the ________________ .
4. *All* of the small islands between the medium-sized island just off the South American coast and

   the Greater Antilles are collectively called the ________________ ________________ .

5. The Lesser Antilles are divided into *two* subgroups known as the ________________ Islands

   and the ________________ Islands.

6. The medium-sized island to the southeast (just off the South American coast) is ____________ .

Answers to the above:

1. Cuba, Hispaniola (or Española), Puerto Rico, Jamaica
2. Florida
3. Bahamas
4. Lesser Antilles
5. Leeward Island (northern half of the group); Windward Islands (southern half of the group)
6. Trinidad

The *thirteen* independent nations of the Atlantic-Caribbean area are:

1. *Bahamas*
2. *Cuba*
3. *Jamaica*
4. *Haiti* (western third of Hispaniola)
5. *Dominican Republic* (eastern two-thirds of Hispaniola)
6. *St. Christopher (Kitts) and Nevis*
7. *Antigua and Barbuda*
8. *Dominica* (most northerly Windward Island)
9. *St. Lucia* (separated from Dominica by French Martinique)
10. *Barbados* (southeast of St. Lucia)
11. *St. Vincent and the Grenadines* (south of St. Lucia)
12. *Grenada* (south of St. Vincent and the Grenadines)
13. *Trinidad and Tobago*

A FEW COMMENTS ON LOCAL USAGE IN THE WEST INDIES

Local inhabitants often use nicknames and abbreviated forms for place-names, and many names are anglicized forms of earlier Spanish names.

Referring to the foregoing list, St. Christopher is usually called St. Kitts; Nevis is pronounced NEE-vis; Antigua is pronounced An-TEEGA; Dominica, Do-min-EE-ka; St. Lucia, St. LOO-sha; Grenada, Gre-NAY-da; Tobago, To-BAY-go.

An interesting local usage in the Bahamas is the use of the name *Caribbean* when referring to the lee (calm) side of an island. Though geographically incorrect, this practice is almost universal in those islands. The *Atlantic side* is an expression reserved for the windward side.

Dependent islands of the West Indies are affiliated with one of the following countries: United States, United Kingdom, France, Netherlands.

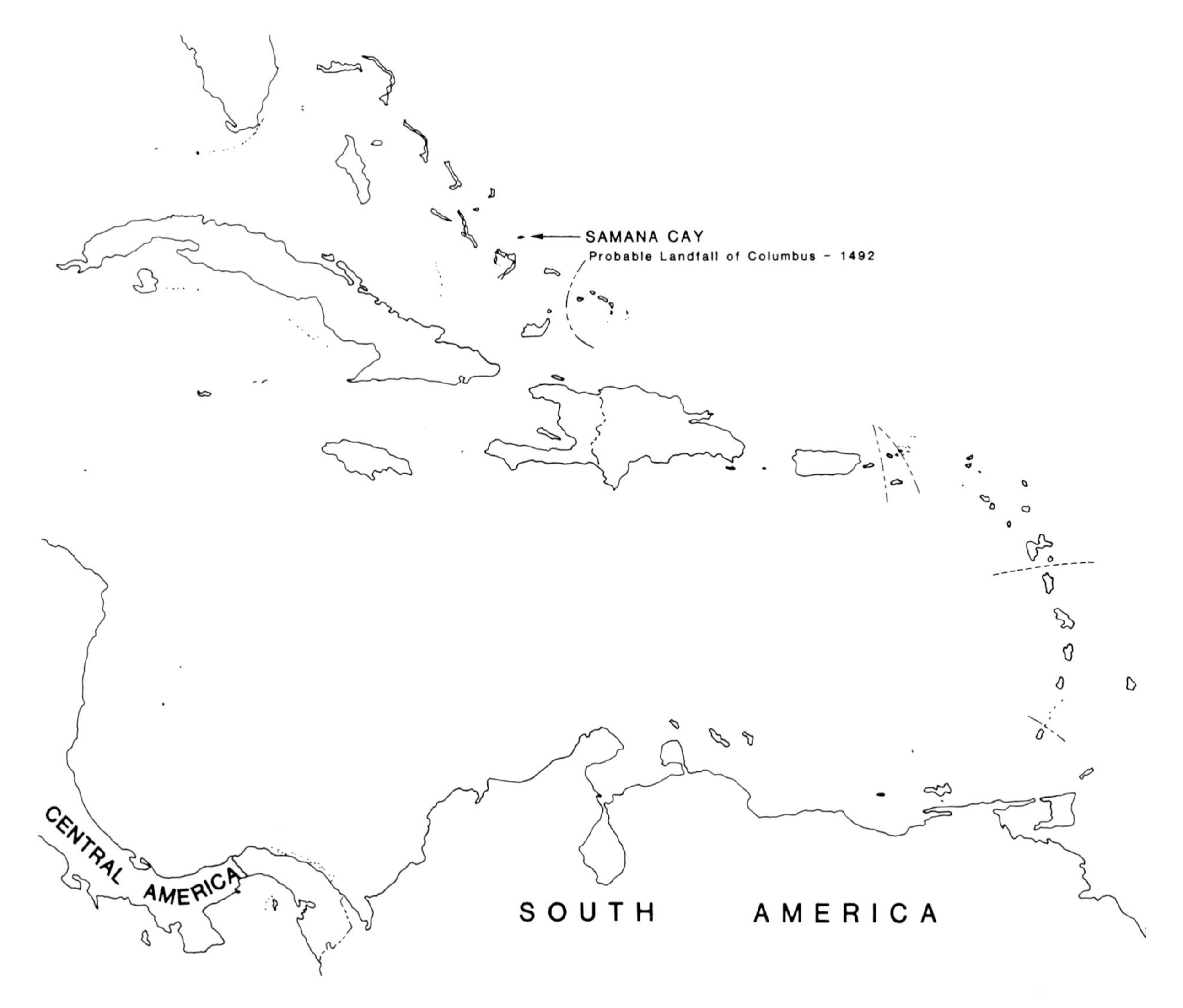

United States:

1. *Puerto Rico:* a free, self-governing commonwealth associated with the United States. Its citizens are also U.S. citizens. Puerto Rico could become the 51st state someday, or possibly independent, depending upon the wishes of its people. *San Juan* is the capital.
2. *Virgin Islands:* 50 islands lying east of Puerto Rico, the most important being *St. John, St. Croix,* and *St. Thomas.* The capital, *Charlotte Amalie,* is on St. Thomas. The Virgin Islands have internal self-government and the people are U.S. citizens.
3. *Navassa:* an uninhabited island between Cuba and Haiti, but the site of a very important lighthouse in the *Windward Passage.*

United Kingdom:

1. *Turks and Caicos Islands:* the last islands in the Bahama chain, but politically a self-governing Associated State of the United Kingdom. *Caicos* is pronounced KAY-kos. The capital is *Cockburn Town,* on *Grand Turk.*
2. *Cayman Islands:* three self-governing islands, east-northeast of Jamaica. The capital is *Georgetown.*
3. *British Virgin Islands:* 36 islands, immediately east of the U.S. Virgin Islands. The capital is *Road Town,* on the island of *Tortola.*
4. *Anguilla* (capital: *The Valley*). The self-governing island is due east of the British Virgin Islands.
5. *Montserrat* (capital: *Plymouth*). A self-governing British island, southwest of Antigua.

France:

All French territories in the Caribbean are administered by either *Guadeloupe* or *Martinique,* each an Overseas Department of France and, therefore, considered integral parts of the French state.

1. *Guadeloupe:* this French Overseas Department lies between Antigua and Dominica. It is the southernmost of the Leeward Islands. The capital is *Basse-Terre.* Islands administered by Guadeloupe include *Marie-Galante, La Desirade,* and *Iles des Saintes* (all located very close to the east and south); *St. Barthélémy* (Bar-te-le-MEE), also called *St. Barts* or *St. Barths* (located north of St. Kitts and Nevis); and *St. Martin* (between Anguilla and St. Barthélémy). This last island is jointly governed with the Netherlands and is the smallest political unit on earth ruled by two countries.
2. *Martinique:* between Dominica and St. Lucia. Its capital is *Fort-de-France.*

Netherlands:

Five islands comprise the *Netherlands Antilles,* and a sixth has affiliation with the group.

1. *Netherlands Antilles:* an entity of the Kingdom of the Netherlands, constitutionally equal with the European homeland. The capital is *Willemstad,* on the island of *Curaçao.* The five-island group is autonomous and self-governing.
   Bonaire (off the coast of Venezeula)
   Curaçao (off the coast of Venezuela)
   Saba (northwest of St. Kitts)
   St. Eustatius (also called *Statia;* northwest of St. Kitts)
   St. Maarten (southern third of island; French northern two-thirds is called *St. Martin;* between St. Eustatius and Anguilla)
2. *Aruba.* In 1986 this island, off the coast of Venezuela and west of Curaçao, became an autonomous member of the Kingdom of the Netherlands, on an equal level with the Netherlands Antilles (to which it formerly belonged) and the Dutch homeland. It still maintains ties with the Antilles as an equal member of the Kingdom. The capital is *Oranjestad.* In many ways Aruba meets the definition of sovereignty and may be considered as such in the near future.

Self-Test

1. Name the thirteen independent states of the Atlantic-Caribbean.

_______________ _______________

_______________ _______________

_______________ _______________

_______________ _______________

_______________ _______________

_______________ _______________

____

2. What is the name of the self-governing commonwealth that is associated with the United States?

    ____

3. What island group is the most easterly territory possessed by the United States in the Western Hemisphere? ____________

4. What European countries have territories in the Lesser Antilles? ____________ ,

    ____________ , and ____________ .

5. What former member of the Netherlands Antilles is now on a governmental par with that organization and is virtually independent? ____________

6. What island is governed jointly by two European states? ____________ What are the governing countries? ____________ and ____________ .

    Check pages 31–33 for the correct answers before continuing.

On the map on page 32 locate (with the help of an atlas) the following cities. Capital cities are in italics.

Nassau, Bahamas
Havana (La Habana), Cuba
Guantánamo, Cuba (a leased U.S. Navy base)
Kingston, Jamaica
Port-au-Prince, Haiti
Santo Domingo, Dominican Republic (oldest city in Western Hemisphere—founded 1496)
Basseterre, St. Kitts and Nevis (the same name as the capital of Guadeloupe, except for the hyphen)
St. Johns, Antigua and Barbuda
Roseau, Dominica
Castries, St. Lucia
Bridgetown, Barbados
Kingstown, St. Vincent and the Grenadines
St. George's, Grenada
Port of Spain, Trinidad and Tobago

A NOTE ABOUT BERMUDA

The Bermuda Islands consist of about 150 coral islands, lying some 920 km (568 mi) east of North Carolina. Only 20 of the islands are inhabited. The main island (Bermuda) is linked to six others by bridges. *Hamilton,* the capital, is on Bermuda. The total population of this self-governing British colony is approximately 60,000. The total area is 53 km² (21 mi²).

Bermuda is NOT in the West Indies, but rather in the open Atlantic. Please avoid this mistake that many make. It is a sad commentary on American geographical knowledge to find some of the 400,000 annual tourists willing to argue with you that Bermuda is in the Caribbean!

3

South America

South America is the large continent adjoining Central America. It is usually considered to be *part* of *Latin America,* though there are many non-Latin portions.

South America consists of *twelve* independent countries and *one* French Overseas Department.

The twelve sovereign nations are:

ATLANTIC COASTAL STATES:	Venezuela
	Suriname
	Guyana
	Brazil
	Uruguay
	Argentina
PACIFIC COASTAL STATES:	Ecuador
	Peru
	Chile
ATLANTIC & PACIFIC STATE:	Colombia
INLAND STATES:	Paraguay
	Bolivia

The French Overseas Department is *French Guiana,* an integral part of the French state. French Guiana and its neighbors, Suriname and Guyana, are often referred to collectively as the *Guianas.*

Using the cartogram on page 36, label the 13 political units of South America. Also identify the Caribbean Sea, Atlantic Ocean, Pacific Ocean, and Central America. (Use the atlas, if necessary.)

A Few Additional Hints:

Bolivia and *Paraguay* are the *only inland* countries in South America. This is easy enough to remember, but *DO NOT* confuse inland Paraguay with coastal Uruguay.

Colombia is quite often misspelled. *There is no "u" in Colombia.*

Ecuador is Spanish for Equator; the country straddles the Equator.

Brazil is the world's *fifth largest country in area* (8,511,965 km² or 3,286,487 mi²) and ranks *sixth* in population (142,000,000). Compare these figures with those on page 7. The name is spelled *Brasil* in Portuguese, the language of the country. *Over half of all South Americans are Brazilians.*

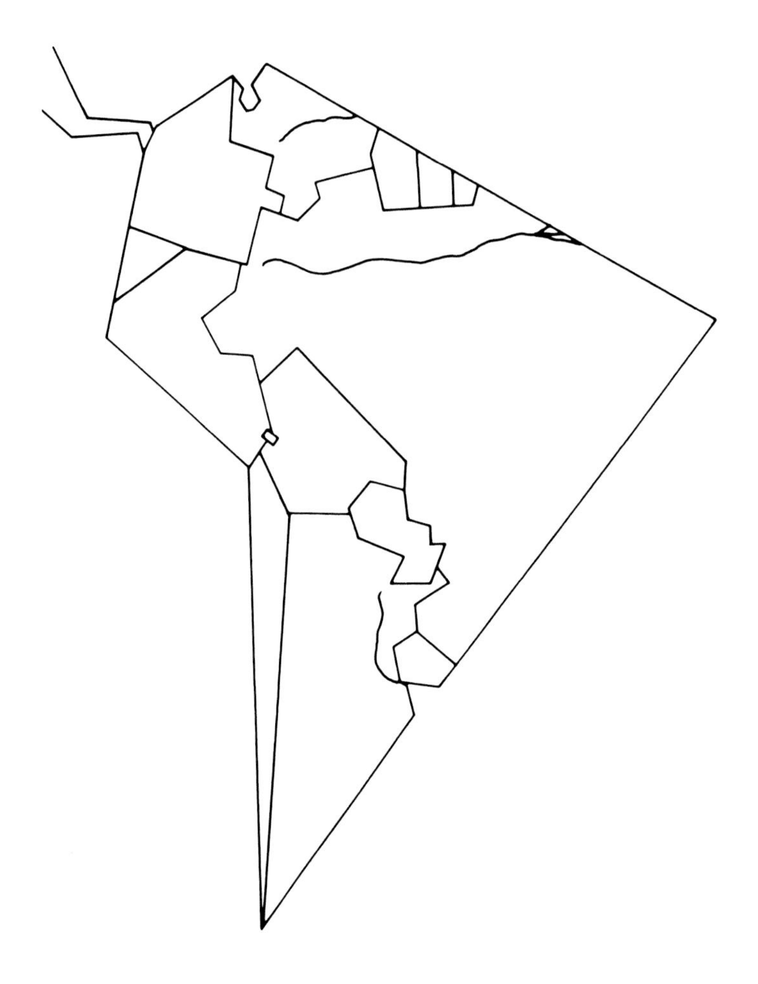

Self-Test

1. What are the only inland states of South America? ______________________

   and ______________________

2. What is the only European-controlled territory in South America?

   ____

3. What is the largest country in South America? ______________________

4. What is the name of the little country on the Equator? ______________________

   ____

5. The small country in South America on the Atlantic coast between Brazil and Argentina is __________

6. ______________________ is a long, narrow Pacific coastal state.

7. ______________ is the country that joins Panama, and fronts both the Pacific and Atlantic coasts.

8. A Pacific coast state between Ecuador and Chile is ______________________

9. The second largest country of South America, located on the Atlantic coast and bordering Chile on the west, is ______________________________

10. Most South Americans speak what language? ______________________________

Check these answers with page 35 *before* continuing.

Label each political unit on the cartogram below. (The dots represent capital cities—forget them for now.)

Two island groups are shown. The one off Ecuador, on the Equator, in the Pacific is the *Galápagos Islands.* They belong to Ecuador. (*Galápago* means "turtle" in Spanish and these islands are famous for their giant sea turtles.)

The islands off Argentina, in the Atlantic, are the *Falkland Islands,* and belong to the United Kingdom. Argentina also *claims* these islands and calls them the *Islas Malvinas* (or Malvinas Islands). The U.K.'s claim dates from the days before the Panama Canal when ships had to sail around the southern tip of South America. It was a coaling stop and the British felt the *Falklands* to be a necessary outpost during the period of undisputed British maritime supremacy. Argentina's claim is based on *proximity.* In 1982 Argentina invaded the Falklands, only to be defeated by the British. Despite its military defeat, Argentina continues to claim the islands.

Note *two lakes* on the cartogram. *Lake Titicaca,* between Peru and Bolivia, is the highest large lake in the world (3,812 m [12,506 ft.]; 8,320 km² [3,200 mi²]). *Lake Maracaibo,* in Venezuela, is the largest South American lake (16,380 km² [6,300 mi²]), though it is not really a lake since it opens to the sea. It might be compared with Lake Pontchartrain in Louisiana, which is also salty, with an outlet to the sea. A major petroleum resource lies beneath Lake Maracaibo's surface and the lake is dotted with oil derricks. About 12 percent of the oil imported by the U.S. comes from here, an area that ranks fifth in world petroleum production.

Label the two lakes and the two island groups on the cartogram below.

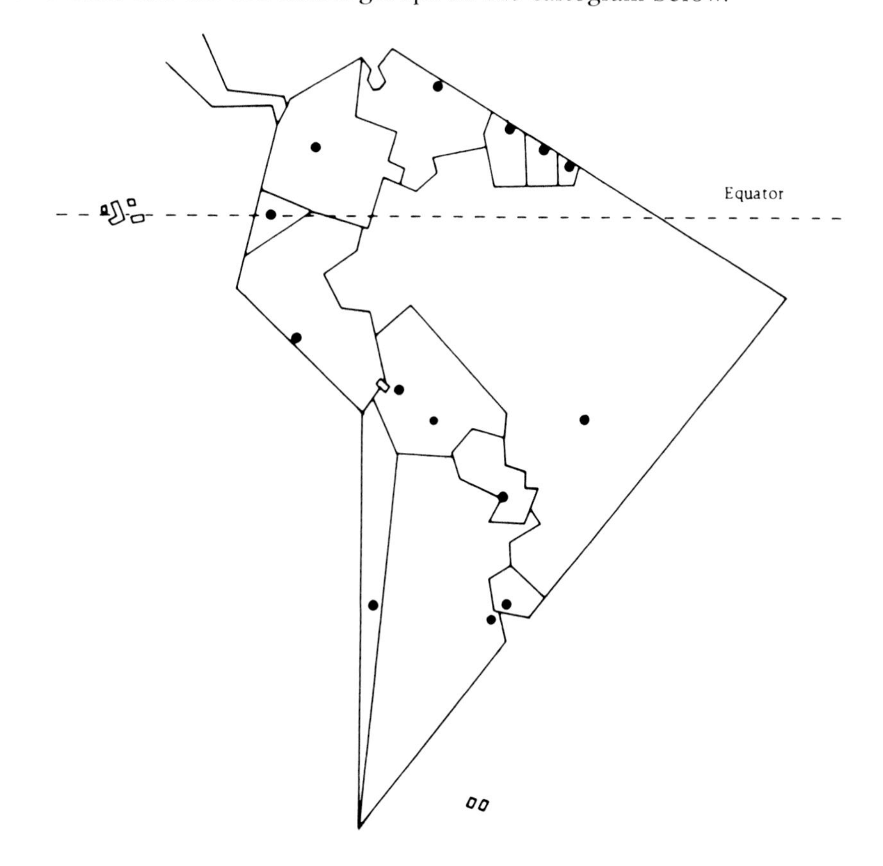

Self-Test

1. What is the name of the island group off the Pacific coast of South America?

   ____

2. What is the name of the island group off the Atlantic Coast?

   ____

3. What country owns the first group? ______________

4. What countries claim the second group? ______________

   ______________ and ______________

5. Name the two largest lakes in South America. ______________

   ______________ and ______________

6. Which lake is famous for petroleum? ______________

7. Which lake is higher? ______________

8. Which is salty and not a true lake? ______________

9. Which lake lies in two countries? ______________

   Name the countries: ______________ and ______________

   ____

10. What country is the larger lake in? ______________

    Check these answers before continuing.

The capitals of the South American countries are:

Venezuela .... *Caracas*
Brazil .... *Brasília*
Uruguay .... *Montevideo*
Argentina .... *Buenos Aires*
Ecuador .... *Quito*
Peru .... *Lima*
Chile .... *Santiago*

Colombia .... *Bogotá*
Paraguay .... *Asunción*
Bolivia .... *La Paz* and *Sucre*
Guyana .... *Georgetown*
Suriname .... *Paramaribo*
French Guiana .... *Cayenne*

Locate each of these cities on the cartogram on page 37. *LEARN EACH!*

Self-Test

Name the countries and capitals of South America.

Country	*Capital*
______________________	______________________
______________________	______________________
______________________	______________________
______________________	______________________
______________________	______________________
______________________	______________________
______________________	______________________
______________________	______________________
______________________	______________________
______________________	______________________
______________________	______________________
______________________	______________________
______________________	______________________

Check these answers and correct any errors before continuing.

This cartogram is to be used with the following three pages.
Refer to the cartogram below throughout this section.

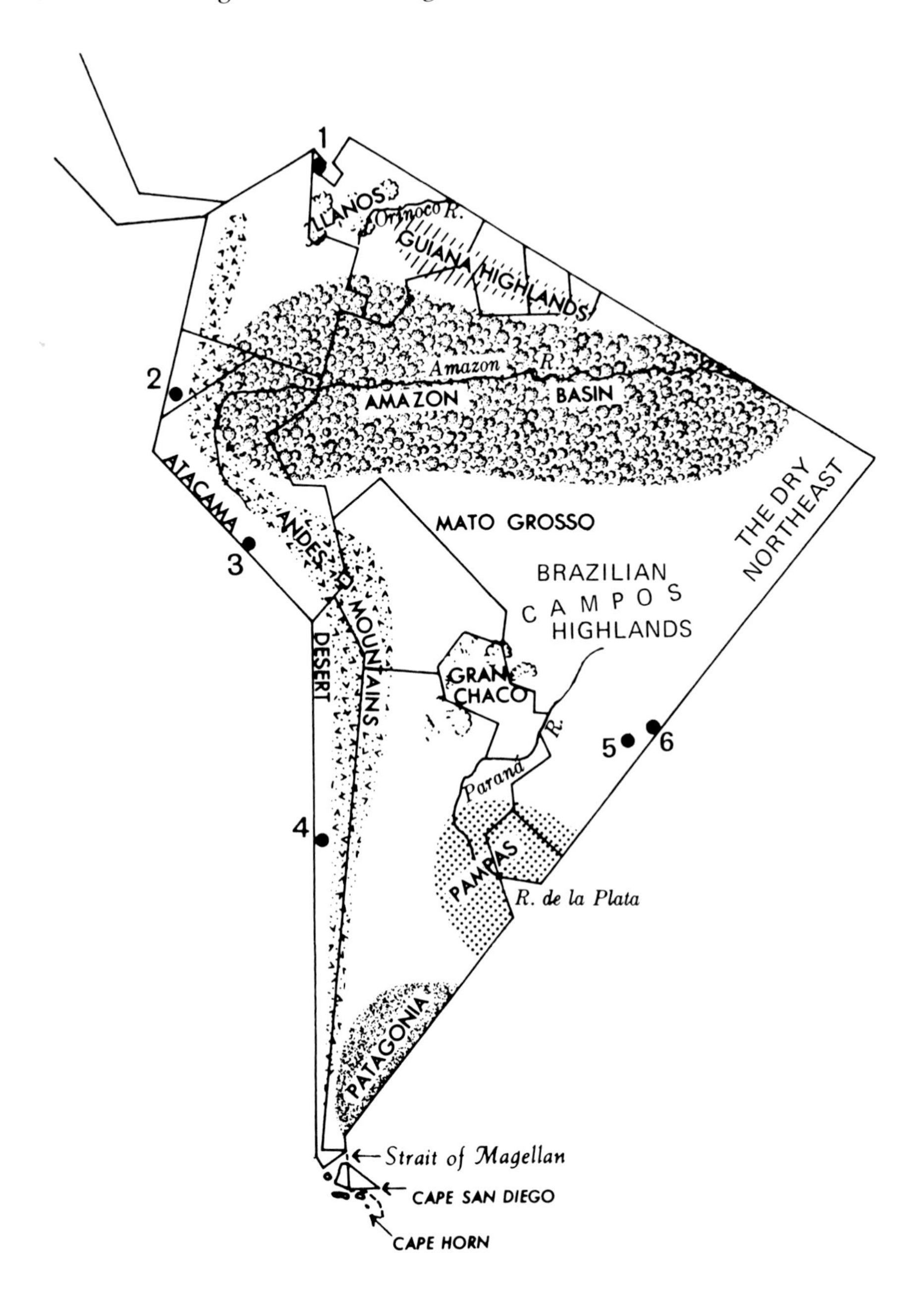

CITIES

1. *Maracaibo,* Venezuela: major oil refining and exporting center on the edge of Lake Maracaibo.
2. *Guayaquil,* Ecuador: major port of that country.
3. *Callao,* Peru: major port of that country.
4. *Valparaiso,* Chile: major port of that country. Note that *all* Pacific coast countries of South America have their capital cities *inland.* There are *no* good natural harbors on the Pacific coast (there are *few* on the west coast of North America!). This is primarily because of geology, where *mountains* come right down to the sea. Also, much of the South American west coast is *desert.*

5. *São Paulo,* Brazil: so-called "Chicago of South America," industrial São Paulo is one of the world's largest cities and is South America's largest (population: ca. 8,000,000).
6. *Rio de Janeiro,* Brazil: former capital and still an important governmental city (population: ca. 5,000,000).

MOUNTAINS AND HIGHLANDS

1. *Andes Mountains:* one of the world's greatest mountain chains, extending from Colombia to Southern Chile and Argentina. At least 50 peaks exceed 6,000 m (20,000 ft.). *Mt. Aconcagua,* Argentina, (6,960 m [22,835 ft.]) is the highest peak in the Western Hemisphere.
2. *Guiana Highlands:* generally in the Guianas, but extending into Venezuela. The highest point is *Mt. Roraima* (2,810 m [9,219 ft.], near the junction of Venezuela—Guyana—Brazil.

   Angel Falls, in the Venezuelan portion of these highlands, is the highest waterfall on earth (979 m [3,212 ft.]).
3. *Brazilian Highlands:* this general descriptive term refers to the hills and mountains of the southern half of Brazil. The escarpment between Rio de Janeiro and São Paulo exceeds 2,700 m (9,000 ft.) in several places.

RIVERS

1. *Amazon River:* rising in the Andes Mountains, 160 km (100 mi) from the Pacific Ocean, the Amazon flows 6,250 km (3,900 mi) to the Atlantic Ocean. Only the Nile River in Africa is longer (6,635 km [4,145 mi]). Its drainage area is called the *Amazon Basin,* which is about two-thirds the size of the United States (or approximately 5,200,000 km² [2,000,000 mi²]).
2. *Orinoco River:* with headwaters in the Guiana Highlands, the Orinoco flows 2,725 km (1,700 mi) through Venezuela to the Atlantic Ocean.
3. *Paraná River:* this is South America's second longest river, flowing 4,000 km (2,500 mi) from the Brazilian Highlands to the Río de la Plata.
4. *Río de la Plata:* this is NOT a river but is an *estuary* formed by the Paraná and Uruguay Rivers.

DESERTS

1. *Atacama Desert:* coastal Peru and Chile. This is one of the driest places on earth. It contains vast deposits of copper and nitrate.
2. *Patagonia:* southern Argentina. This is a cold, bleak desert. It is one of only two *east* coast deserts on earth (the other is in east Africa).

GRASSLANDS AND PLAINS

1. *Llanos:* a tree-studded grassland (savanna) in Venezuela and Colombia. The word is Spanish for *plains* and is applied to many similar areas throughout Spanish America.
2. *Mato Grosso:* literally "tall grass" in Portuguese. A scrubby, hilly savanna in west central Brazil.
3. *Campos:* Portuguese and Spanish for "fields." This is the savanna of southern Brazil and includes the *Mato Grosso.*
4. "*The Dry Northeast*": the bulge of northeastern Brazil, which consists of extremely dry savanna. This is one of South America's major economic problem-areas.
5. *Gran Chaco:* thorn forest, or dry savanna, mainly in Paraguay.
6. *Pampas:* the lush, tall-grass prairies of northeast Argentina. The *pampas* are one of the world's richest livestock and agricultural areas. They are the home of the *gaucho,* or Argentine cowboy.

Tierra del Fuego ("Land of Fire") is the large island at the southern tip of South America, separated from the continent by the *Strait of Magellan.* It was named by Magellan in 1520 for the many Indian campfires he saw.

Label this island on the preceding cartogram (page 40).
Study the small outline map that follows.

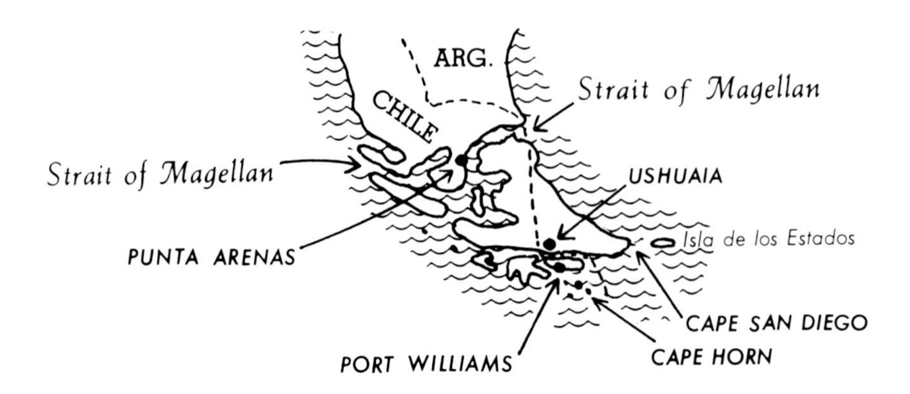

Cape Horn is not a true geographic cape as is *Cape San Diego,* but rather it is an *island.* The name *Horn* is from a Dutch town called *Hoorn.* It is only coincidental that *Tierra del Fuego* is shaped like a horn and this causes most people to misplace the Cape. Note also the international boundary between Argentina and Chile.

This is the end of the section on South America. At the same time, this completes a study of the place-names of the entire Western Hemisphere. It might be worth the effort to pause for a moment before moving on and review the entire book up to this point. Certainly, it would be wise to review South America and to ask questions from the last few pages on cities, mountains, rivers, plains, and deserts. Study the atlas, and locate additional items that have not been covered, such as climates and vegetation.

Before concluding this portion, there is one question that should have occurred and which needs answering. Just what is Latin America? It is that part of the Western Hemisphere where the people speak a language of the Latin branch (usually Spanish, Portuguese, and French). Normally, French-speaking Canada is NOT included as part of Latin America, but it is more properly a part of it than some Indian areas and non-Latin regions south of the Rio Grande. Latin America most often means that area south of the United States but excluding all those areas where English or Dutch is the predominant language. It does include areas where Indian languages dominate, though this is more for convenience than for accuracy.

When ready, turn the page and continue.

4

Europe

Europe is a politically complex area of *34 sovereign states* (NOT counting the USSR). These should be learned *thoroughly* at the outset, preferably in logical groupings.

I. Fenno-Scandinavian Countries*

1. Iceland
2. Denmark
3. Norway
4. Sweden
5. Finland

Study the arrangement of these countries in the atlas and on the cartogram on page 54. Note that *Norway extends north of Sweden and Finland* and has a common boundary with the USSR.

Territorial possessions: Remember, *Greenland* belongs to Denmark. The *Faeroe Islands,* located between Iceland and Norway, is also a Danish possession. North of Norway, in the Arctic Ocean, lies the Norwegian possession of *Svalbard (Spitzbergen).*

When *certain* in the knowledge of the *five* Fenno-Scandinavian countries, continue.

II. The British Isles

There are *two* sovereign states in the British Isles:

1. The United Kingdom of Great Britain and Northern Ireland (often called the United Kingdom or the U.K.)
2. The Republic of Ireland

Study the map carefully.

The four units that comprise the United Kingdom should not be thought of as "states." Their present status is the product of hundreds of years of political evolution. *England* developed from two kingdoms (Angle-Danish and Saxon-Jute), but was not really unified until the Norman conquest in A.D. 1066. *Wales* became a principality in A.D. 1301, and since that time England and Wales have been administered as one unit. *Scotland* was a separate kingdom until A.D. 1603. In that year there was a "union of crowns," when King James VI of Scotland became King James I of England. Nevertheless, Scotland retains some autonomy and a legal system based on Roman law. *Northern Ireland* has only been a part of the United Kingdom since 1921.

* *Fenno-Scandinavian* is a geological term and refers to the ancient crystalline rock formation (called a ***shield***) that extends from Finland through Norway. *Scandinavia* is the peninsula that contains Sweden and Norway. Inasmuch as the cultures of Norway, Denmark, and Iceland have a common origin, and that of Sweden is closely related, the countries are usually treated as a unit.

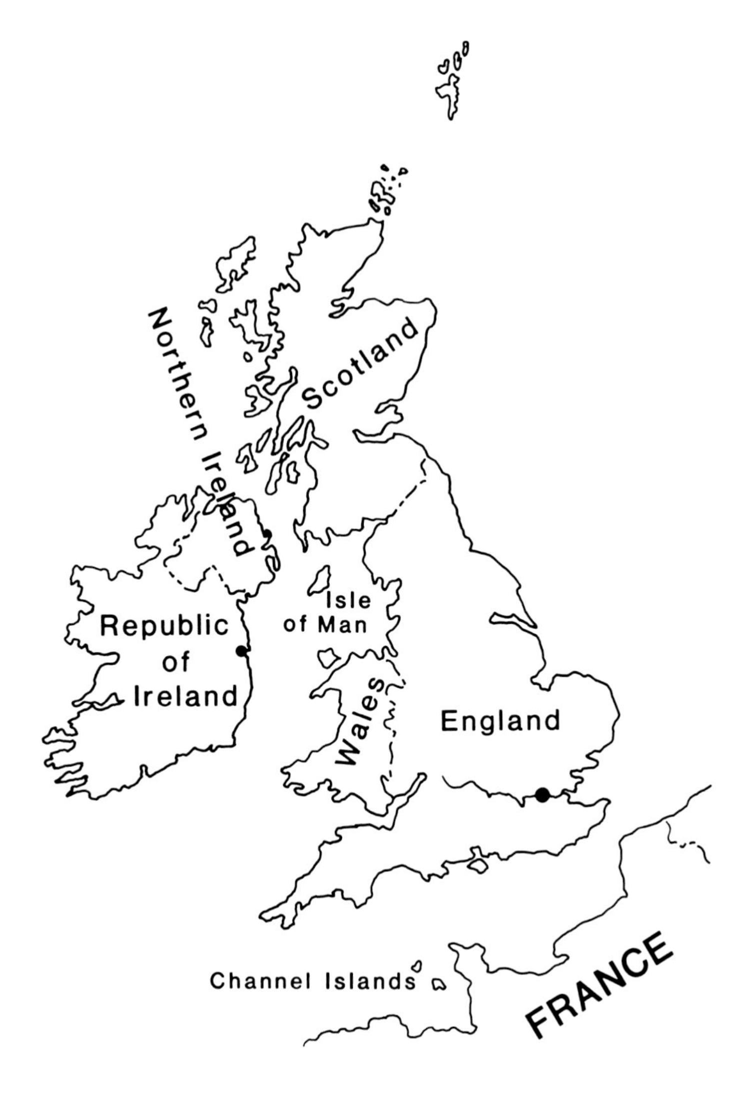

Two Crown dependencies of the United Kingdom (*but not part of it*) are the *Channel Islands* and the *Isle of Man*. Each has a Lieutenant Governor appointed by the Crown and an elected Parliament. The Isle of Man lies between England and Northern Ireland; the Channel Islands, just off the French coast, are west of Normandy. The best known of the Channel Islands are *Jersey* and *Guernsey*.

The name *Great Britain* means *either* (1) the name of the larger of the two main British Isles, or (2) the collective units of *England, Scotland,* and *Wales*. The unit is often called *Britain*.

The name *Ireland* should be used only for the smaller of the two principal British Isles, but in common practice it is used to refer to the *Republic of Ireland. Northern Ireland* is not politically affiliated with the *Republic of Ireland*.

Remember: England + Wales + Scotland + Northern Ireland = *The United Kingdom. The Republic of Ireland* stands alone as a separate, independent nation.

Self-Test

Name the *five* Fenno-Scandinavian countries.

__________________, __________________, __________________, __________________, ____________

The United Kingdom is composed of *four* political subunits:

__________________, __________________, __________________, __________________

The other sovereign nation in the British Isles is ____________________________________

__________________ which is also known as __________________

Check these answers with pages 43 and 44 before continuing.

Refer to the atlas. On the map on the preceding page, indicate and label the following:

1. London
2. Dublin
3. Shetland Islands
4. Orkney Islands
5. Outer Hebrides
6. Irish Sea
7. English Channel
8. Strait of Dover
9. North Sea

East of Scotland, approximately 160–320 km (100–200 mi), lie some of the most important oil fields in the world. Located in the north-central portion of the North Sea, these fields have made the U.K. self-sufficient in natural gas and petroleum. While this may pull the U.K. out of an economic nose dive, it is certain to create new problems. Recent Scottish demands for more local autonomy (and even independence) are related directly to the newfound wealth.

Self-Test

The *northernmost* of the British Isles will soon possess the largest oil port in Europe.

Name the islands. __________________

In what *sea* is the oil found? __________________

The oil is located east of __________________

How far offshore is the oil? __________________ (km) __________________ (mi)

Check your answers above.

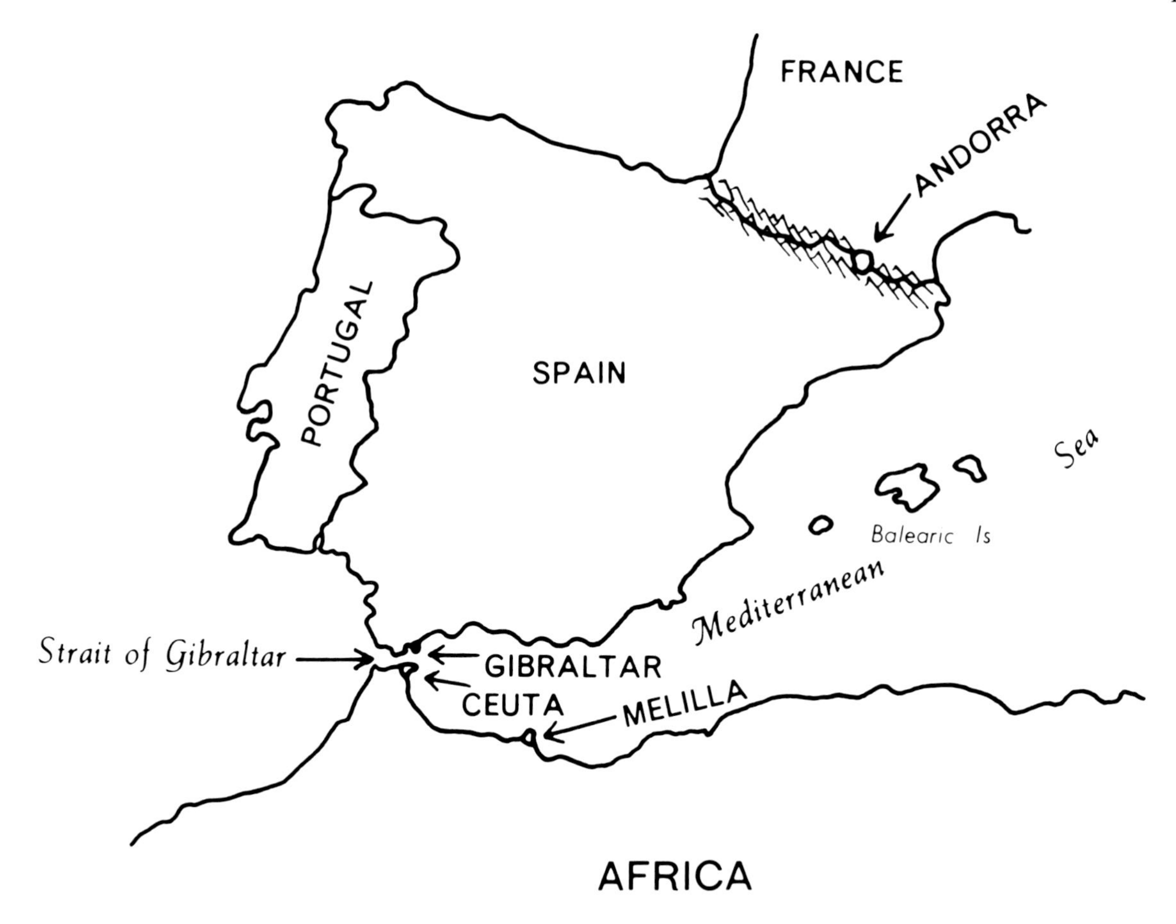

III. Iberian Peninsula

The Iberian Peninsula is composed of:

1. Portugal
2. Spain

Ceuta and *Melilla* in Africa, and the *Balearic Islands* are considered integral parts of Spain. *Gibraltar* is a *British* colony on the Spanish mainland and is NOT an island. *Andorra,* in the *Pyrenees Mountains,* is one of six microstates in Europe and will be discussed later.

Both Portugal and Spain possess islands in the Atlantic Ocean (*not shown on the map*) that are integral parts of each state.

The *Canary Islands* are part of Spain. They lie approximately 1,250 km (800 mi) southwest of Spain. The nearest of the seven principal islands is only 80 km (50 mi) from the African mainland. Columbus sailed to America from the Canary Islands. The name has nothing to do with a bird of the same name, but is derived from the Latin word *canae* (dogs).

The *Madeira Islands* (1,000 km [600 mi] southwest of Portugal) and *Azores* (1,600 km [1,000 mi] west of Portugal) are provinces of Portugal. *Look these up in your atlas.*

IV. The "Little Three"—Benelux

Three of Western Europe's smallest countries are *Belgium, Netherlands,* and *Luxembourg.* Collectively, they are often referred to as the *Benelux* countries (a word formed by using the first syllable from each name). Often they are called the *Low Countries* (meaning low in elevation). This term is not really a good geographic term, except for the *Netherlands* (which actually means "low lands"), and the northern half of Belgium, but it is commonly used. Remember *Benelux* (pronounced Ben-e-lux): *Be*lgium, *Ne*therlands, *Lux*embourg.

Locate the Benelux countries on the map below.

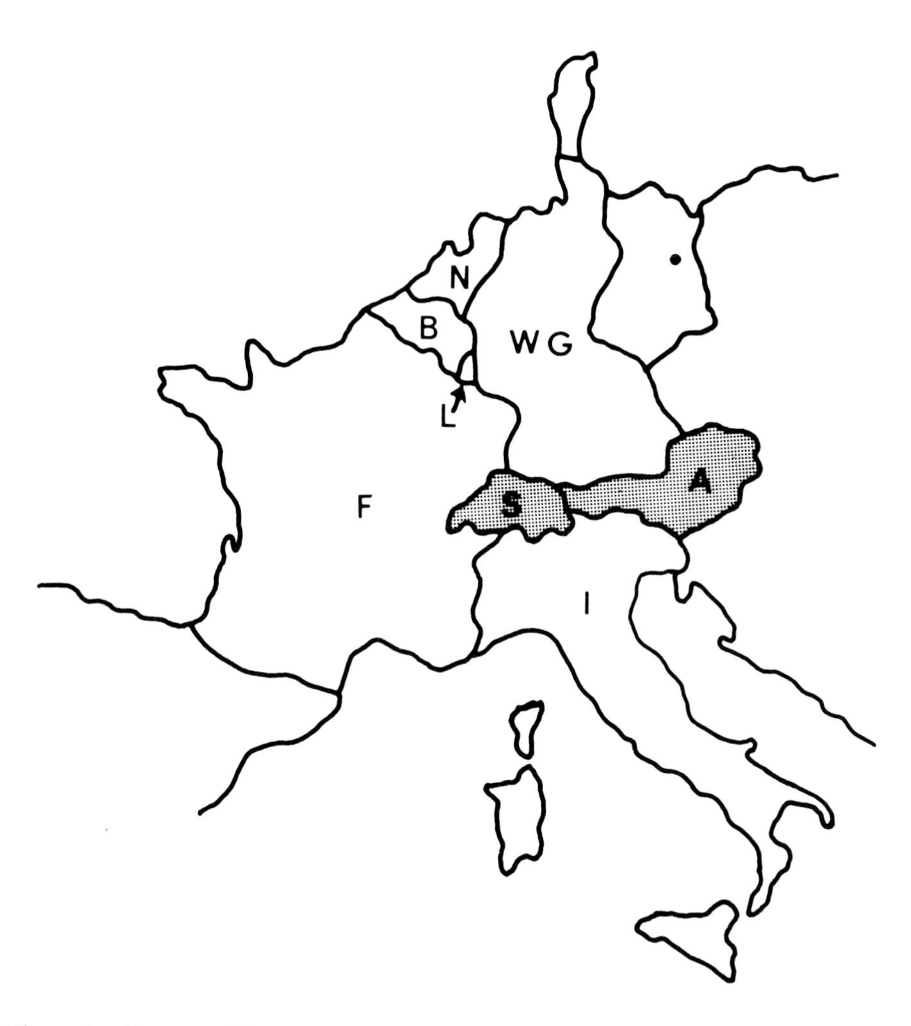

V. The "Big Three"—France, West Germany, and Italy

Continental Europe's "Big Three" (excluding, of course, European USSR) are *France,* the *Federal Republic of Germany,* and *Italy.*

Locate these on the map above. Also, label the three large Mediterranean islands: *Corsica* (the smallest) belongs to France; *Sardinia* (just south of Corsica) and *Sicily* (off the toe of the Italian boot) are Italian.

Note: The *Federal Republic of Germany* is commonly called *West Germany.* Though one should know the proper term, it is seldom used in colloquial speech. The *German Democratic Republic* is *East Germany. East Germany* should NOT be confused with *West Germany.* Get in the habit of always using the prefix *West* or *East* with Germany, since there are now *two* Germanys. The map (following) should clear up any confusion concerning *West* and *East Germany.*

A WORD ABOUT THE EUROPEAN ECONOMIC COMMUNITY ("COMMON MARKET")

In 1958, the six states named in IV. and V. (above) formed an economic union. In the ensuing years six additional states became members (United Kingdom, Republic of Ireland, Denmark, Greece, Portugal, and Spain). The twelve countries now comprising the European Economic Community have a combined population of about 325 million people and form one of the world's major economic units.

A WORD ABOUT BERLIN

The city of Berlin lies in the heart of East Germany—a capitalist island in a communist sea. At the end of World War II, Berlin was occupied by the USSR, France, the United Kingdom, and the United States. The Soviet zone became *East Berlin.* The French-British-American zone became *West Berlin.* Although the western portion is a state within the Federal Republic of Germany (West Germany), the city is still technically occupied and some restrictions still apply. West Berlin is a detached piece of West Germany, lying 160 km (100 mi) from the rest of the country. The infamous Berlin Wall makes for a vivid and symbolic boundary between democratic West Berlin and the surrounding communist territory.

Carefully study the map below before continuing.

Self-Test

1. Name the five Fenno-Scandinavian countries:

_______________, _______________, _______________, _______________,

____

2. Name the two sovereign states of the British Isles:

_______________, _______________

3. Name the two countries of the Iberian Peninsula:

_______________, _______________

4. Name the very small state (microstate) found in the Pyrenees Mountains between France and Spain:

____

5. Name the Benelux countries:

__________, __________, __________

6. Name the "Big Three" of continental Western Europe:

__________, __________, __________

Check these answers with the preceding pages before moving on.

VI. Switzerland and Austria

These two mountainous nations, lying partially in the *Alps Mountains,* are often thought of together. They act as a wedge between Italy and West Germany. *Switzerland* and *Austria* are shaded on the map that appears on page 47.

VII. The Microstates of Europe

There are six countries in Europe that are so small that they are referred to as *microstates.* Mention has already been made of *Andorra,* one of the six.

The European miniature states are:

Microstate	*Area*	
	km²	*mi²*
Andorra	487	188
Malta	316	122
Liechtenstein	160	62
San Marino	61	24
Monaco	01.5	00.6
Vatican City	00.4	00.17

Though extremely small (the largest is only 23 km [14 mi] long by 23 km wide!), each of these six states fits the definition of statehood. They are self-governing, independent nations. Some date from medieval days, others (like Malta) are twentieth century creations. Vatican City and Monaco are so small that together they could fit into New York City's Central Park with room to spare!

Vatican City has worldwide influence as the center for the Roman Catholic Church. The Pope is the chief-of-state.

Monaco is the site of the famous Monte Carlo Casino.

All six states are well known to stamp collectors inasmuch as stamps provide a significant portion of each country's revenue. Often, they are called Europe's "postage stamp" nations.

Study the maps on page 50 and locate each nation on page 54.

Andorra is on the map on page 46.

Note: SMOM (the Sovereign Military Order of Malta) is sometimes considered to be the world's smallest country. Located in downtown Rome, SMOM occupies one office building and does enjoy extraterritorial status. In many respects SMOM meets the definition of statehood.

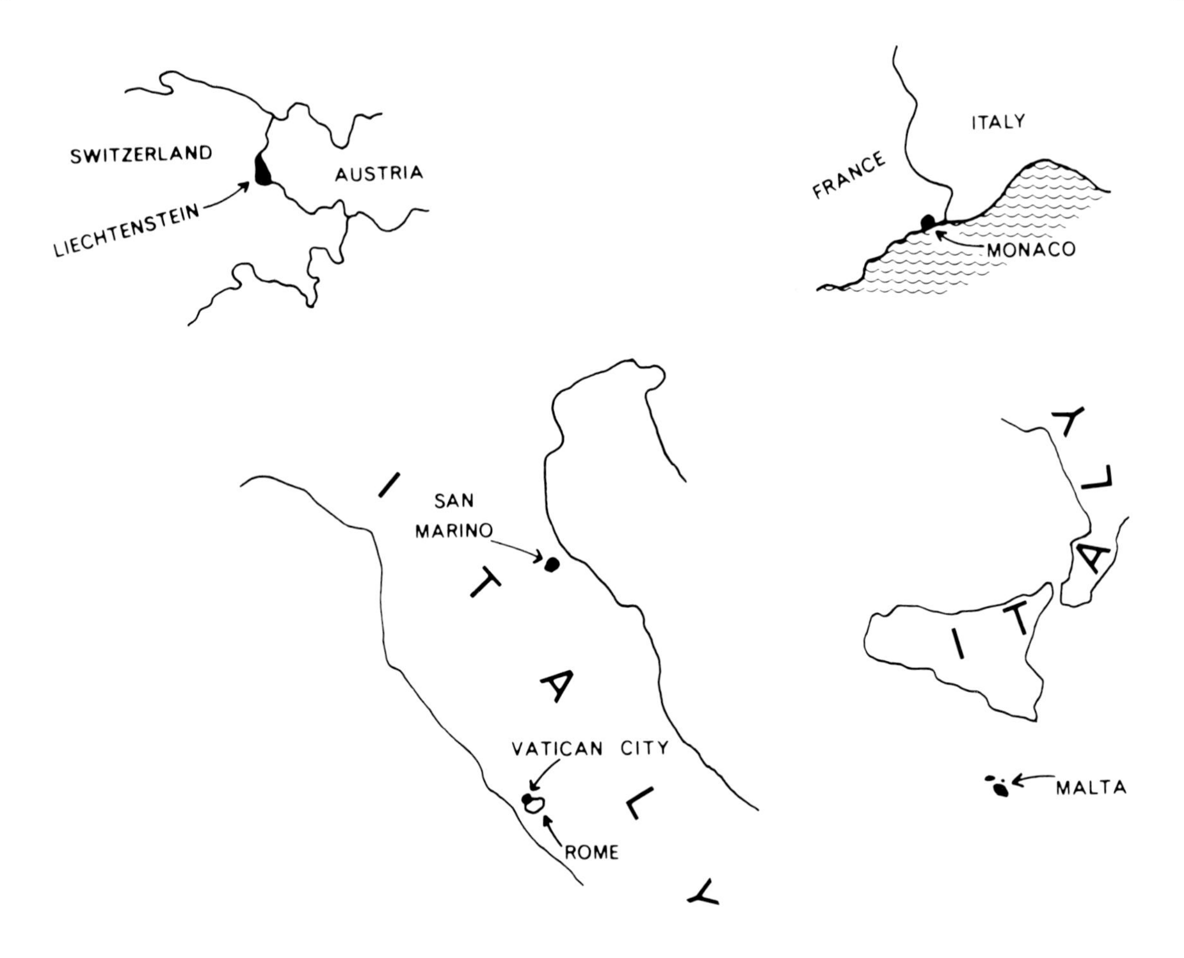

Summary of European Countries to this Point:

5 Fenno-Scandinavian:	Iceland
	Denmark
	Norway
	Sweden
	Finland
2 British Isles:	United Kingdom
	Republic of Ireland
2 Iberian Peninsula:	Portugal
	Spain
3 "Little Three":	Belgium
	Netherlands
	Luxembourg
3 "Big Three":	France
	West Germany
	Italy
2 Alpine:	Switzerland
	Austria
6 Microstates:	Andorra
	Malta
	Liechtenstein
	San Marino
	Monaco
	Vatican City

23 TOTAL TO DATE

Eleven countries still remain to be considered. *Three* are in the eastern Mediterranean; *eight* are Eastern European states.

VIII. The Eastern Mediterranean States

1. Greece
2. Turkey
3. Cyrpus

Greece, Turkey, and *Cyprus* lie at the eastern end of the Mediterranean Sea. Turkey is only *partly* in Europe proper, and Cyprus is an island. Cyprus, situated just south of Turkey, has a population that is four-fifths Greek and one-fifth Turkish. Cultural conflict has turned into open warfare on several occasions, with Greece supporting the Greek Cypriot majority and Turkey taking sides with the Turkish Cypriot minority. In 1983 the Turks made a unilateral declaration that created the *Turkish Republic of Northern Cyprus.* To date it has gone unrecognized by the international community

Study the map, following.

The *shaded* portion is that part of *Turkey in Europe.*

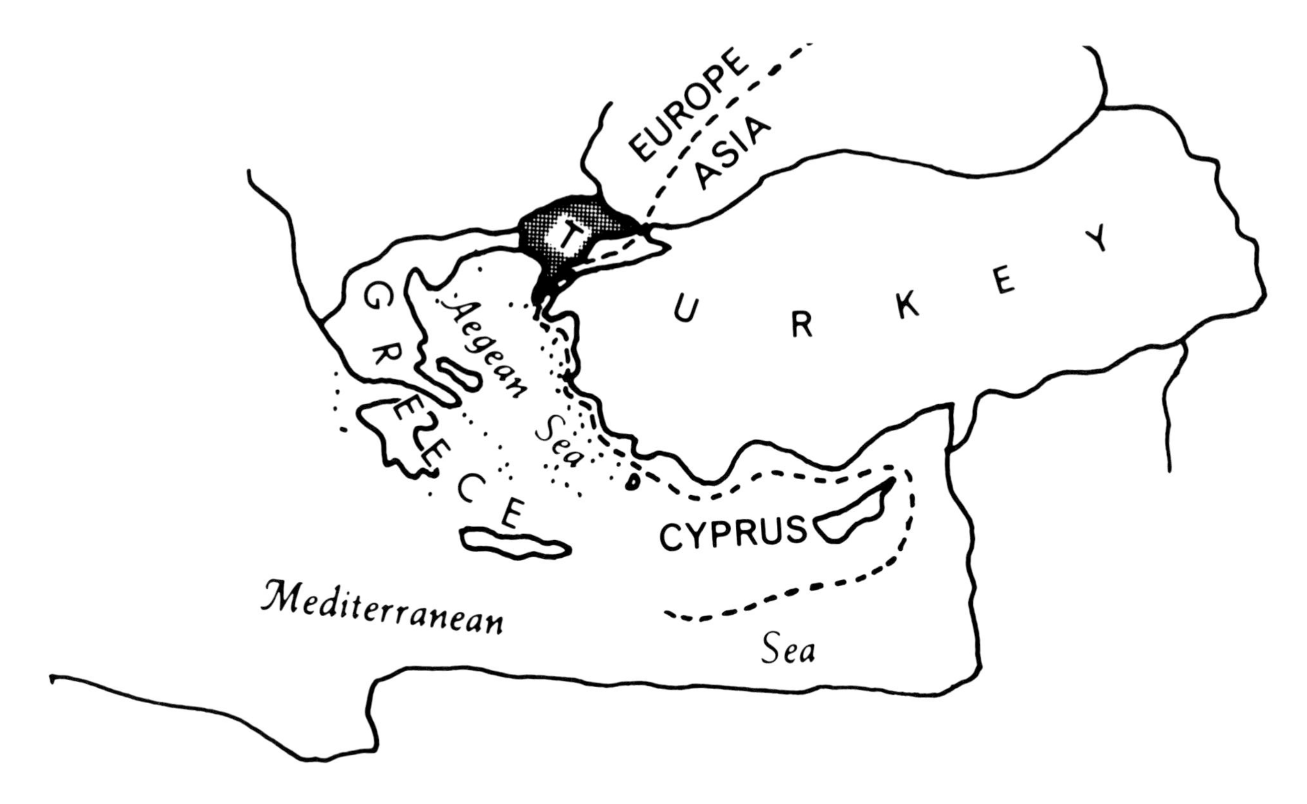

Note that Greece is largely composed of peninsulas and islands. The Greek islands just off the Turkish coast are in *Europe;* the Turkish mainland (often only a mile [1.6 km] or so away) is in *Asia.*

Asiatic Turkey is also called *Asia Minor* or *Anatolia,* when a regional (non-political) term is used. The interior of Asiatic Turkey is a plateau, called the *Anatolian Plateau.*

Study below the detail of the straits connecting the *Black Sea* and the *Mediterranean Sea.* Turkish control of this area has long been a thorn in the side of the Soviets. Turkey, a western ally (and NATO member), could easily seal off the Black Sea in a few hours.

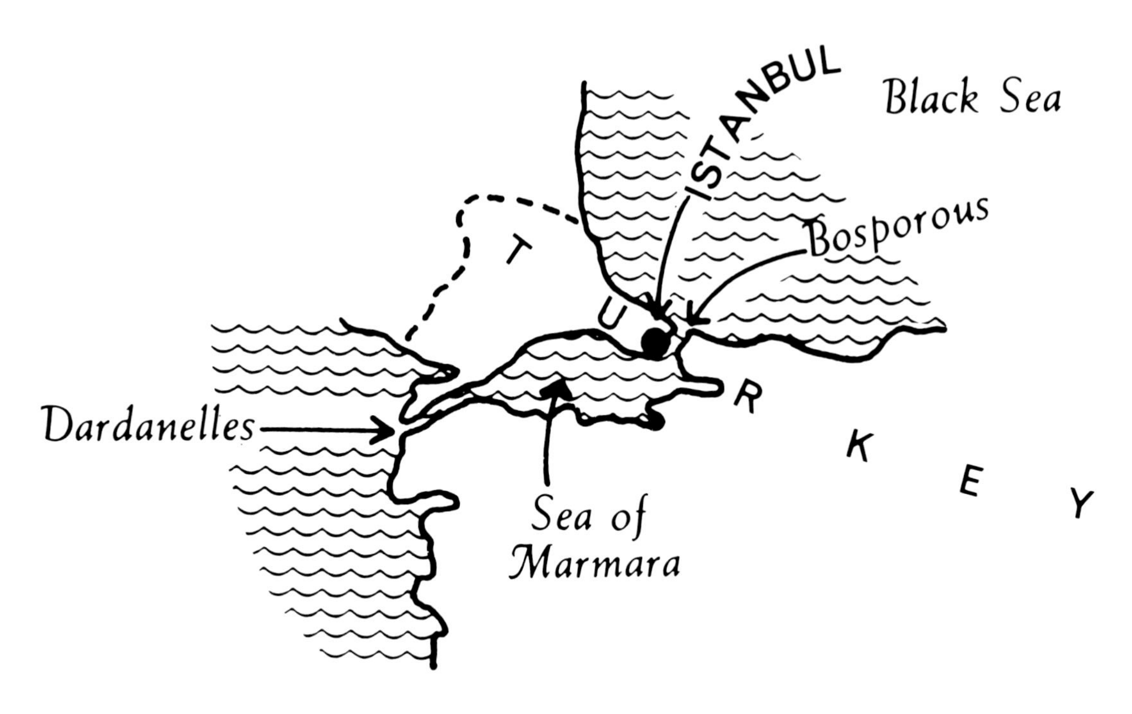

IX. Eastern Europe

A zone of *eight* communist countries separates the countries of Western Europe from the Soviet Union.

(Ideologically, Greece, Turkey, and Cyprus are *Western* countries.)
Study the cartogram on page 53 and use the list above to help label each country indicated.

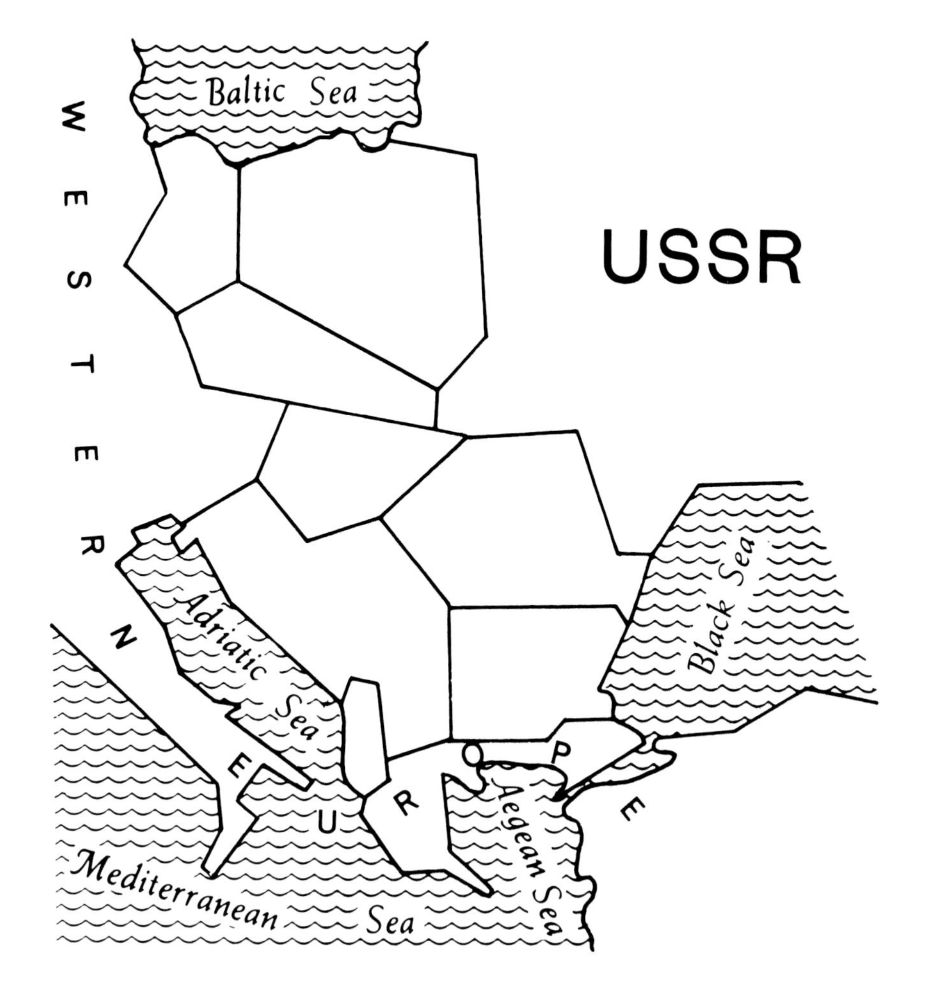

Label the 34 sovereign countries of Europe on the following cartogram (excluding the USSR).
Label the *one* British colony in continental Europe.
Label the following water bodies:

1. Mediterranean Sea
2. North Sea
3. Baltic Sea
4. Black Sea
5. Adriatic Sea
6. Aegean Sea
7. Ionian Sea
8. Tyrrhenian Sea
9. Bosporous
10. Dardanelles
11. Sea of Marmara
12. Strait of Gibraltar
13. Bay of Biscay
14. English Channel
15. Irish Sea
16. Skagerrak
17. Kattegat
18. Gulf of Bothnia
19. Gulf of Finland
20. Atlantic Ocean

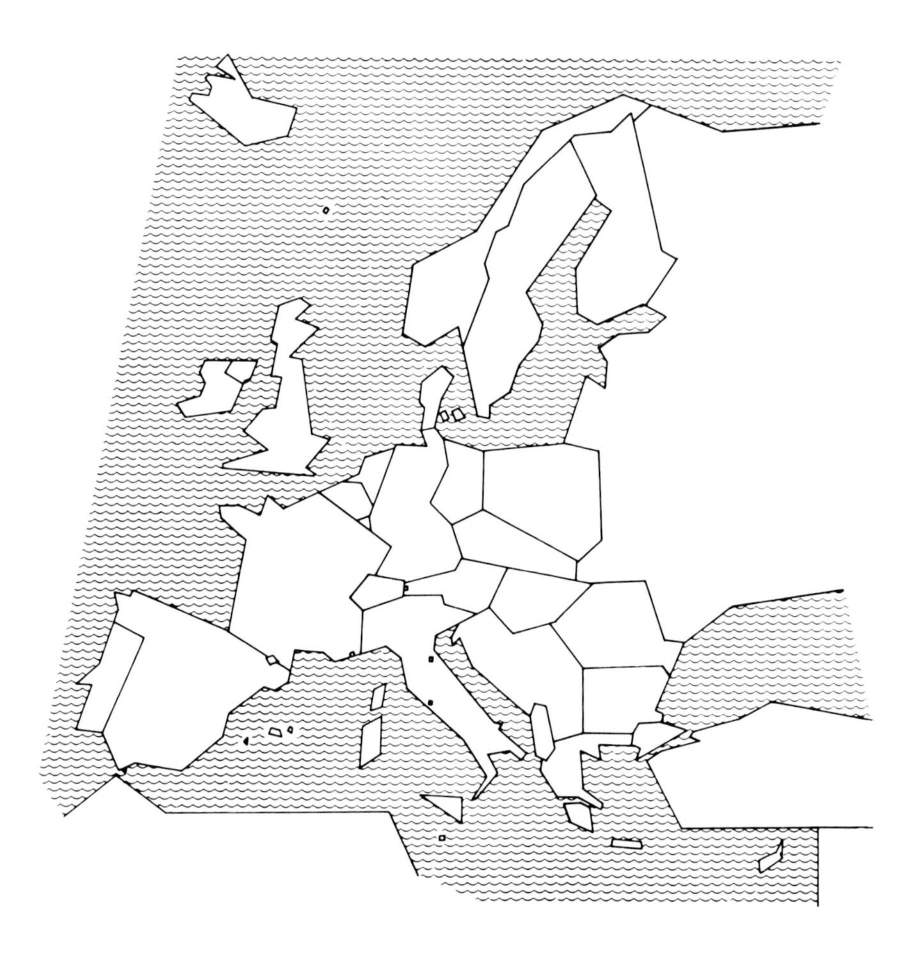

Omit the six microstates for this exercise.

Name the other 28 sovereign countries of Europe with the help of the atlas and give the capital city for each.

	Country	*Capital*
Fenno-Scandinavian:	____________	____________
	____________	____________
	____________	____________
	____________	____________
	____________	____________
British Isles:	____________	____________
	____________	____________

Iberian Peninsula: ________ ________

________ ________

"Little Three": ________ ________

________ ________

________ ________

"Big Three": ________ ________

________ ________

________ ________

Alpine: ________ ________

________ ________

Eastern Mediterranean: ________ ________

________ ________

________ ________

Communist Countries: ________ ________

________ ________

________ ________

________ ________

________ ________

________ ________

________ ________

________ ________

Locate *each* capital city named above on the cartogram on page 56. (Omit the six microstates.)

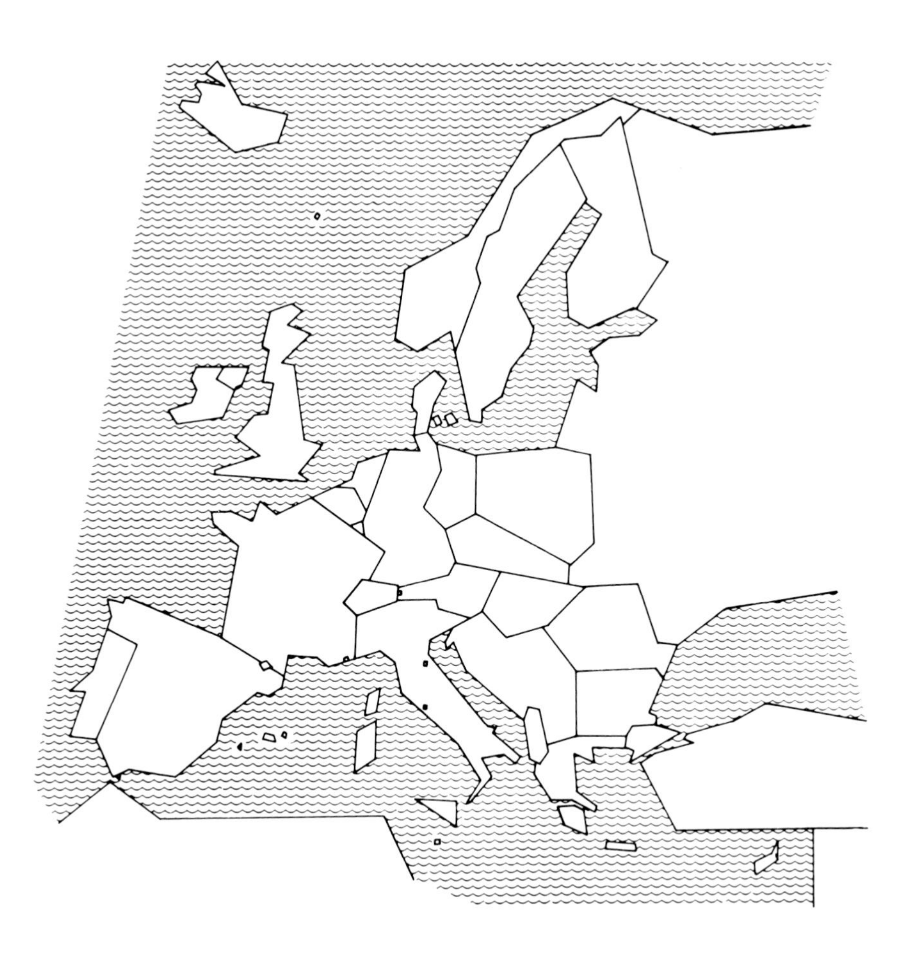

The *common English names* for the capitals of Europe are given below. Names in parentheses are those of a local language or of a secondary language spoken in the particular country.

Country	*Capital*
Iceland	*Reykjavik*
Denmark	*Copenhagen (København)*
Norway	*Oslo*
Sweden	*Stockholm*
Finland	*Helsinki*
United Kingdom	*London*
Republic of Ireland	*Dublin (Baile Átha Cliath)*
Portugal	*Lisbon (Lisboa)*
Spain	*Madrid*
Belgium	*Brussels (Brussel, Bruxelles)*
Netherlands	*The Hague ('s-Gravenhage)* and *Amsterdam*
Luxembourg	*Luxembourg*
France	*Paris*
West Germany	*Bonn*
Italy	*Rome (Roma)*
Switzerland	*Bern*
Austria	*Vienna (Wien)*

Greece .... *Athens (Athínai)*
Turkey .... *Ankara*
Cyprus .... *Nicosia (Levkosia)*
East Germany .... *East Berlin*
Poland .... *Warsaw (Warszawa)*
Czechoslovakia .... *Prague (Praha)*
Hungary .... *Budapest*
Romania .... *Bucharest (Bucureşti)*
Yugoslavia .... *Belgrade (Beograd)*
Albania .... *Tirana (Tiranė)*
Bulgaria .... *Sofia (Sofiya)*

DO NOT CONTINUE UNTIL YOU KNOW THE CAPITALS OF EUROPE!

The number of European cities (even large ones) is almost infinite, and knowledge of them all is not expected. It is strongly recommended, however, that everyone look up the larger ones in the atlas. A few are mentioned below and on the next page and special note should be made of these.

Major Cities:

(in order of countries as given on pages 56 to 57, where appropriate)

Sweden .... Göteborg
United Kingdom:
England .... Birmingham, Liverpool, Manchester, Sheffield
Northern Ireland .... Belfast
Scotland .... Glasgow, Edinburgh
Wales .... Cardiff
Spain .... Barcelona, Bilbao, Seville (Sevilla)
Belgium .... Antwerp (Antwerpen)
Netherlands .... Rotterdam
France .... Bordeaux, Le Havre, Lyon, Marseille
West Germany ....Cologne (Köln), Hamburg, Munich (München)
Italy .... Milan (Milano), Naples (Napoli)
Switzerland .... Geneva (Genève), Zürich
Greece .... Salonika (Thessaloníki)
Turkey .... Istanbul
East Germany .... Leipzig, Dresden
Poland .... Gdańsk, Lódź

USE THE ATLAS TO FIND EACH OF THESE AS MENTIONED.

Industrial Regions: (with major city and location)

English Midlands	Birmingham	northwest England
Ruhr Basin	Essen	northwest West Germany
Po Valley	Milan (Milano)	north Italy

Plains: (important agriculture)

North European Plain
Paris Basin
Hungarian Basin
Wallachian Plain .... Romania

Rivers:

Thames .... England
Ebro .... Spain
Po .... Italy
Elbe .... Czechoslovakia, East & West Germany
Oder (Odra) .... Czechoslovakia, Poland, East Germany
Vistula (Wislá) .... Poland
Rhine (Rhein) .... Switzerland, West Germany, France, Netherlands

Danube (Donau, Duna, Dunaj, Dunav, Dunarea) .... West Germany, Austria, Czechoslovakia, Hungary, Yugoslavia, Bulgaria, Romania

Garônne .... France
Rhône .... Switzerland, France
Saône .... France
Loire .... France
Seine .... France
Tagus (Tajo, Tejo) .... Spain, Portugal

Mountains:

Alps .... France, Italy, Switzerland, Austria
Pyrenees .... Spain, France
Dinaric Alps .... Yugoslavia
Carpathians .... Czechoslovakia, USSR, Poland, Romania
Transylvanian Alps .... Romania
Pindus .... Greece
Apennine .... Italy
Pennine .... England (U.K.)

Note: The highest mountains in Europe are the *Caucasus,* in the USSR. (Mt. Elbrus: 5,633 m [18,481 ft.])

Other Locations:

Saar .... West Germany (coal)
Lorraine .... France (iron ore)
Silesia .... Poland (iron, coal)

Kiruna .... Sweden (iron)
Scottish Lowlands .... Scotland, U.K. (coal)
Ardennes .... Belgium (coal)
Ploeşti .... Romania (petroleum)

The above lists are not all-inclusive. They are intended only to show a *few* of the major factors of European place geography. Study Europe very carefully in the atlas, including maps of landforms, vegetation, climate, and economics. Knowing the countries and principal features is a necessary first to understanding a most highly developed and complex region.

5

Union of Soviet Socialist Republics (USSR)

The Union of Soviet Socialist Republics is the world's largest country in area (22,272,200 km² [8,599,300 mi²]), and with a population of 285,000,000, it ranks only behind China and India in number of people. Stretching across half of Europe and Asia, the USSR covers one-sixth of the earth's total land surface.

The USSR is a political union of *fifteen* union republics, "republics," or, more properly, "Soviet Socialist Republics." The "republics" of the USSR are comparable in many ways to the "states" of the United States, and, like the United States, the USSR is a federal state (that is, certain powers reside with the subordinate political units).

The Soviet Socialist (union) Republics may be grouped as follows:

I. *One* giant *Eurasian* republic that dominates all others
II. *Six* republics on the *western* margins of the USSR
III. *Three southern* republics, between the Black and Caspian Seas
IV. *Five Asiatic* republics

Study the cartogram below. The fifteen republics are numbered from 1 to 15, and these numbers are used in the discussion of each republic as a means of identification.

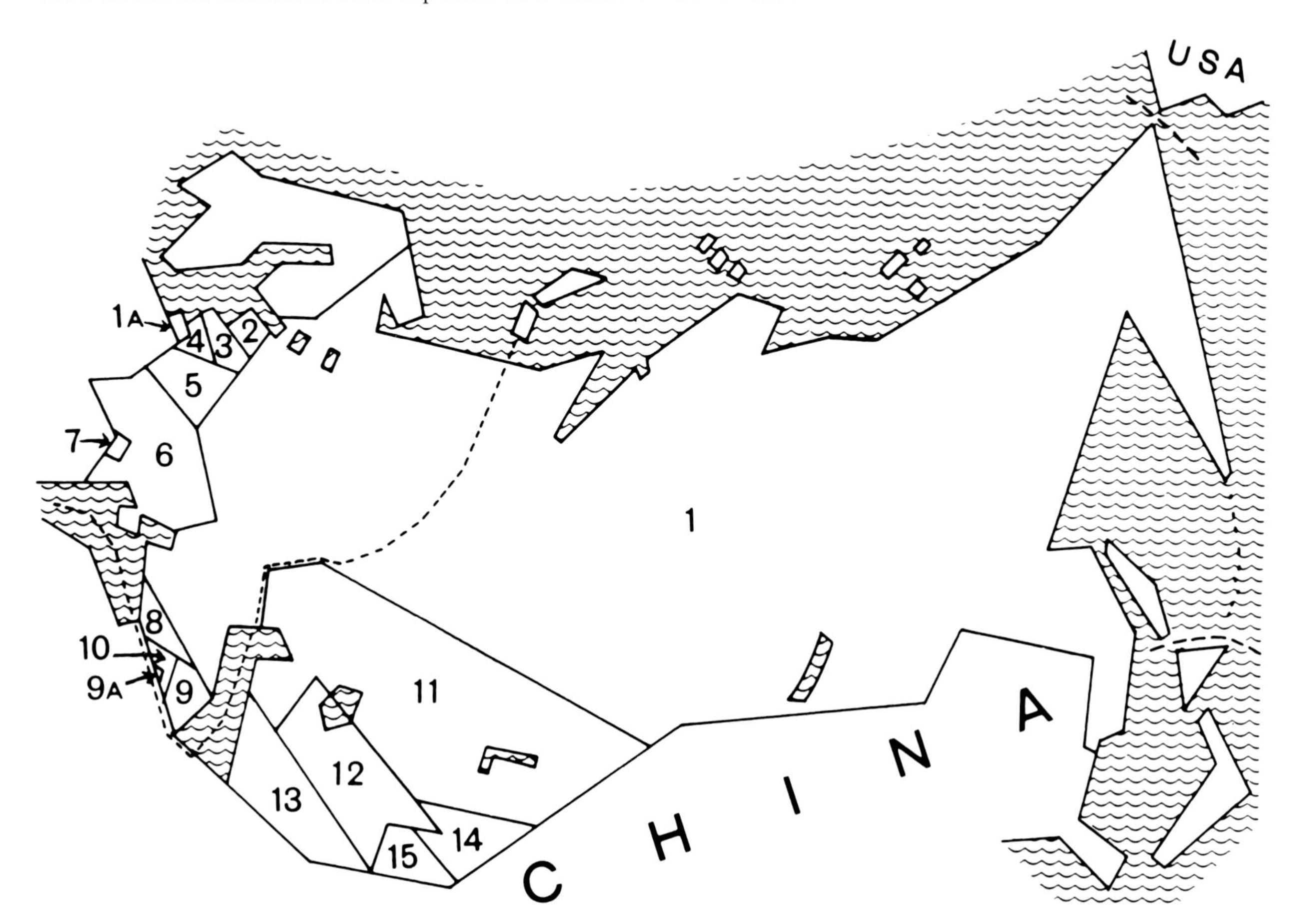

I. The Largest Soviet Republic

1. The *Russian Soviet Federated Socialist Republic* (more easily called RSFSR) is a vast geographic area (almost *twice* as large as the United States) and has over 50 percent of the entire population of the USSR. This republic may be thought of as *Soviet Russia* proper. It extends across Europe and Asia and is the *only* one of the fifteen Soviet Socialist Republics to do so.

Note the dotted line on the cartogram (page 60). This is the boundary between Europe and Asia, which cuts the RSFSR in two.

Also, note the little region numbered 1a. This is a *detached* portion of the RSFSR. The Soviet Union does this in one other case (see nos. 9 and 9a), a rare practice among federal states.

II. The Six Western Soviet Socialist Republics

From north to south, as numbered on the cartogram, the six western Soviet Socialist Republics are:
Offical Name:

2. Estonian Soviet Socialist Republic
   Common Name—Estonia
3. Latvian Soviet Socialist Republic
   Common Name—Latvia
4. Lithuanian Soviet Socialist Republic
   Common Name—Lithuania
5. Belorussian Soviet Socialist Republic
   Common Name—White Russia
6. Ukrainian Soviet Socialist Republic
   Common Name—Ukraine
7. Moldavian Soviet Socialist Republic
   Common Name—Moldavia

Estonia, Latvia, and *Lithuania* were independent states *before* the USSR incorporated them into the union after World War II. Together, they are often called the *Baltic States,* since they face the *Baltic Sea.* All six of these western Republics are in Europe.

Self-Test

1. Name the largest republic of the USSR:

   ____

2. Name the six western republics of the USSR:

   ____

   ____

   ____

   ____

   ____

   ____

Check these answers before continuing.

III. The Three Southern Republics

Lying between the *Black Sea* and the *Caspian Sea* and in Europe's highest mountains, the *Caucasus Mountains,* are three Soviet Socialist Republics.

Official Name:

8. Georgian (Gruziya) Soviet Socialist Republic
   Common Name—Georgia
9. Azerbaydzhan Soviet Socialist Republic
   Common Name—Azerbaydzhan
10. Armenian Soviet Socialist Republic
    Common Name—Armenia

Georgia is easy to remember inasmuch as it is the name of a "deep south" state in both the USSR and the United States.

Armenia is fairly well known because of the Armenian people, many of whom have come to the United States in the past. It should be noted that many Armenians live across the border in *Turkey.* The Soviets have gotten a lot of propaganda mileage out of the fact that the Armenians in Turkey are a minority group and the USSR claims Turkish oppression of these people. The Turkish-Soviet Armenian border is a rather tense area. Recently, Armenia has claimed territory within Soviet Azerbaydzhan.

Azerbaydzhan is a tribal name that is not familiar to most Americans. The *Azerbaydzhani* have their own language and customs (as is true in most Soviet Republics). The name is pronounced *Azer-by-john.* They are also a minority in Iran. The population is Muslim.

Note the little area numbered 9a. This is a *detached* portion of *Azerbaydzhan.*

IV. The Five Asiatic Republics

Five Soviet Republics are *totally* within *Asia.* Three of them are largely desert and scrubby grassland; two are deep in the mountains bordering *China* and *Afghanistan.*

These remote republics are inhabited primarily by Asiatic peoples—many racially *Mongoloid* and of the *Muslim* religion. It is an area of camels, oases, mosques, and typically "dry world" culture for the most part. Sometimes, the entire region is referred to as *Russian Turkestan, West Turkestan,* or *Turan.* Similar peoples live across the border in China, where the region is often called *Chinese Turkestan* or *East Turkestan.* Along this border there has been open conflict between the USSR and the People's Republic of China.

The five Asiatic Republics are:

11. Kazakh Soviet Socialist Republic (Kah-zahk)
12. Uzbek SSR (Ooze-beck)
13. Turkmen SSR
14. Kirgiz SSR (Keer-giz)
15. Tadzhik SSR (Tah-jik)

These may seem like strange names, but they each identify a tribal or ethnic group.

The term *Cossack* is dervied from *Kazakh,* though the former became a social group and the latter was (is) an ethnic group. The Soviet equivalent of Cape Canaveral is located in *Kazakh,* and all Soviet Cosmonauts have been launched from there.

Uzbek possesses the USSR's finest cotton lands and is a major region of irrigation.

Turkmen, inhabited by Turkic peoples (the Turks of *Turkey* came originally from this general area and speak a closely related language), is four-fifths desert.

Kirgiz and *Tadzhik* are mountainous lands of great natural beauty.

IF there were one more republic in this area that had a name beginning with the letter "U," one could simply remember

K U T   K U T—but, in reality, it must be:
K U T   K   T.
Think about the rhythm here:
*K*azakh, *U*zbek, *T*urkmen, *K*irgiz, *T*adzhik

Review the *fifteen* republics of the USSR before going on. When you are certain that you know them *all* (and can spell each one *correctly!*), cover the top of this page and take the self-test.

Self-Test

Name the fifteen Soviet Socialist Republics.

LARGEST: 1. ______

WESTERN: 2. ______

3. ______

4. ______

5. ______

6. ______

7. ______

SOUTHERN: 8. ______

9. ______

10. ______

ASIATIC: 11. ______

12. ______

13. ______

14. ______

15. ______

CHECK! Are all fifteen correct? DO NOT PROCEED UNTIL SURE OF ALL.

USA

Use the atlas to complete the following. *Label* the map on page 64. All items to be named are indicated on the map.

Cities:	*Pronunciation*
Alma-Ata	Alma-Ahta
Astrakhan (Astrachan)	Ah-strah-khan
Baku	Bah-koo
Dnepropetrovsk	Nep-ro-peh-trofsk
Frunze	Froon-zeh
Gorki (Gor'kij)	Gorky
Irkutsk	Ear-kootsk
Kiev (Kijev)	Kee-ev
Leningrad	Lenin-grahd
Minsk	Minsk
Moscow (Moskva)	Moscow
Murmansk	Mur-mansk
Novosibirsk	Novo-sih-bersk
Odessa	O-dess-ah
Riga	Ree-ga
Rostov	Ros-tov
Samarkand	Sam-ar-kand
Sverdlovsk	Sverd-lovsk
Tashkent (Taskent)	Tash-kent
Verkhoyansk (Verchojansk)	Ver-ko-yansk
Vladivostok	Vlad-i-vos-tock
Volgograd	Vol-go-grad

Rivers:

Amur	Ah-mur
Dnieper (Dnepr)	Knee-per
Don	Don
Lena	Lee-na
Ob	Ohb
Syr Darya (Syrdarja)	Seer Dar-ya
Volga	Vohl-ga
West Dvina	Dvee-na
Yenisey (Jenisej)	Yen-i-say

Mountains:

Ural	Your-al
Caucasus	Kaw-ka-sus
Pamir-Tien (Tyan) Shan ranges	Pa-meer Tee-en

Seas and Lakes:

Aral Sea	Air-al
Lake Baikal (Bajkal)	By-call
Lake Balkhash (Balchas)	Ball-cash

Baltic Sea	
Bering Sea	
Black Sea	
Caspian Sea	
East Siberian Sea	
Sea of Japan	
Kara Sea	Care-ah
Lake Ladoga	Lad-o-ga
Laptev Sea	Lap-tev
Sea of Okhotsk	Ok-hote-sk
White Sea	

Other Locations:

Kuznetsk Industrial Region (Kuzbass)	Kooze-netsk (Kooze-bahs)
Donetsk Basin (Donbass)	Don-etsk (Don-bahs)
Novaya Zemlya	No-va-ya Zeml-ya
Valdai Hills	Val-die
Sakhalin Island	Sak-ha-leen
Kurile Islands	Koor-eel
Japan	
China	
Korean Peninsula	
United States (Alaska)	

CERTAIN INTERESTING POINTS

1. Connect *Leningrad, Odessa,* and *Novosibirsk* by drawing a triangle. Approximately 90 percent of the agriculture, population, industry, and stock raising of the USSR lies within this *triangle.*
2. *Little Diomede Island, U.S.,* lies 3.9 km (2.4 mi) from *Big Diomede Island, USSR.* The international boundary runs *between* these islands, and the U.S. and USSR do, in fact, touch at this point. In Russian, *Big Diomede* is called *Ostrov Ratmanova.*

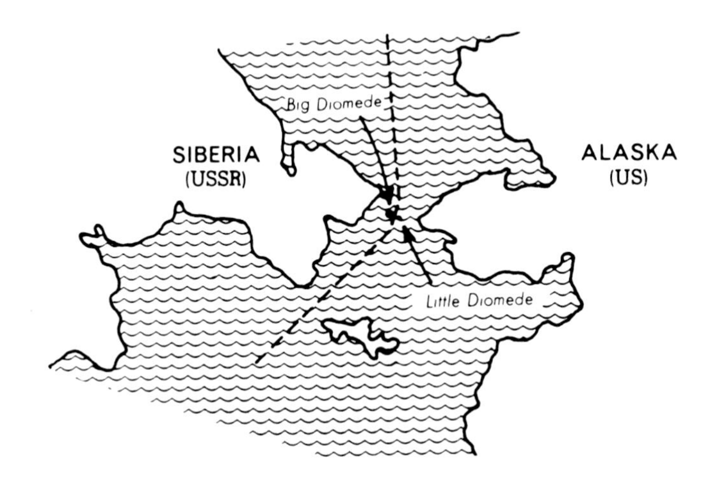

3. The area between the *Ural Mountains* and the *Yenisey (Jenisej) River* is the *West Siberian Lowland.* When the spring thaw comes, this is the world's largest swamp.
4. The *Ob* (5,567 km [3,460 mi]), *Amur* (4,344 km [2,700 mi]), *Lena* (4,312 km [2,680 mi]), *Volga* (3,685 km [2,290 mi]), and *Yenisey (Jenisej:* 3,347 km [2080 mi]) rivers are among the world's largest. The *Volga* is the longest river in Europe. Only the Nile and Amazon exceed the *Ob* in length.

5. *Lake Baikal* (Bajkal) is the *world's deepest lake* (1,620 m [5,315 ft.]), with the greatest volume (23,000,000 m^3 [812,237,410 $ft.^3$]). Ten percent of *all* fresh water on earth is contained by Lake Baikal.
6. The *Caspian Sea* is technically a *lake* (completely surrounded by land), and is therefore the *world's largest lake*—almost *five times* as large as Lake Superior (U.S./Canada). The *Aral Sea* is also a lake, ranking *fourth* largest on earth. Both the Caspian and Aral are salt water.
7. *Novaya Zemlya* was the site of the world's largest nuclear explosion, which took place just before the U.S./USSR Nuclear Test-Ban Treaty became effective October 10, 1963. This Arctic Ocean Island has been a primary Soviet nuclear test area.
8. *Siberia* is actually the *Asiatic portion of the RSFSR.* Generally everything east of the Ural Mountains and north of the desert (Kazakh Republic) would be considered *Siberia.*

This is the end of the USSR portion. Review it carefully before continuing with the next chapter.

6

Asia

Asia—that vast area of *Eurasia* lying east of the Ural Mountains and Mediterranean Sea—may be divided a number of ways for study. *Soviet Asia* has already been examined in the preceeding section. *Turkey,* the only other country besides the USSR that lies *both* in Europe and Asia, was considered earlier as a part of Europe.

The *three* principal subdivisions of Asia (excluding the USSR and Turkey) that will be discussed in the following pages are:

I. Southwest Asia
II. Southern and Southeast Asia
III. East Asia

These three realms of Asia are indicated on the outline map that follows.

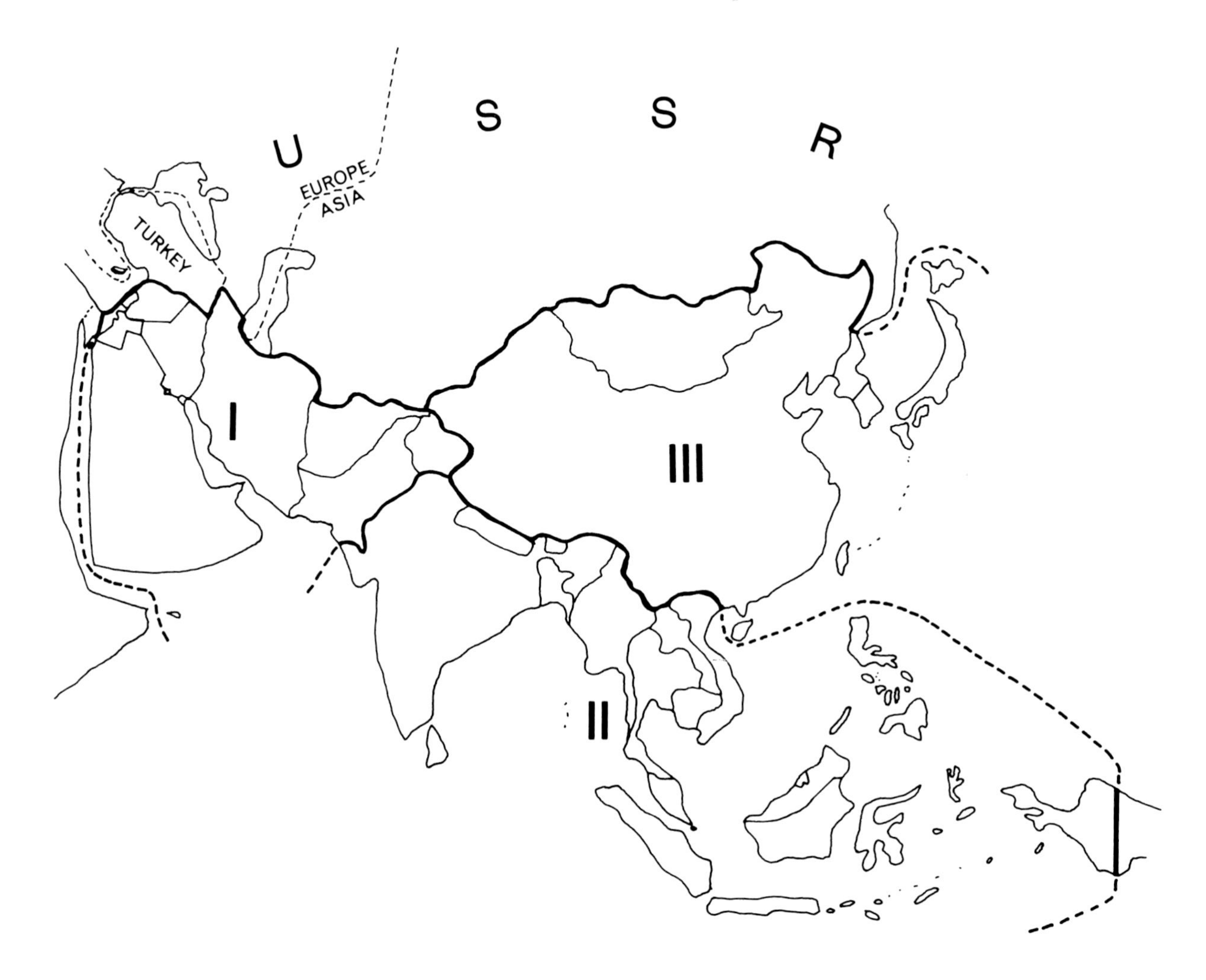

I. Southwest Asia

Southwest Asia is a complex political realm. Much of the territory is *desert* and is occupied by tribes that traditionally have paid little attention to formal political boundaries. Vaguely defined areas are ruled by Arab *sheiks (shayks)* in some cases, with various degrees of autonomy.

Refer to the cartogram on the next page while following this discussion.

With the exception of the Jewish state of Israel, all of the countries of this realm are *Muslim*. Lebanon, however, has a large Christian minority, and Israel, a large Muslim minority.

Southwest Asia is often called the *Middle East*. This rather general term was coined by the British during World War II and has many meanings. *Turkey* is almost always included in the term *Middle East*, and *Egypt* often is. It would *probably* be correct to say that Southwest Asia, as defined on the cartogram by the numbered countries, is the Middle East *minus* Turkey and Egypt.

Though most of Southwest Asia is Muslim, it is *NOT* entirely *Arab*. The *Arab World* is that part of the Middle East *and* North Africa where most people speak the Arabic language.

Pakistan (1), *Afghanistan* (2), and *Iran* (3) are *non-Arab* Muslim countries. *Turkey* is also a *non-Arab* Muslim country. Israel, of course, is non-Arab *and* non-Muslim.

Again, with Israel as the exception, the countries *south* and *west* of Turkey and Iran are Arab states. A glance at a good physical (topographical) map will show that the area from Pakistan to Turkey is a high plateau. *This plateau is non-Arab country*. The Arabs live south and west of this plateau.

The term *Near East* usually means the *Arab* region shown plus Turkey and Israel.

One must be very careful when using imprecise terms like *Middle East, Near East,* and even *Far East*. They mean different things to different people.

See below for the key to numbers on the cartogram on page 70.

Southwest Asia consists of:

16 independent states (numbers 1–16)
1 disputed territory (17)
1 neutral zone (18)

Independent States:

1. Pakistan
2. Afghanistan
3. Iran (formerly called "Persia")
4. Iraq (formerly called "Mesopotamia")
5. Syria
6. Lebanon
7. Israel
8. Jordan
9. Kuwait
10. Saudi Arabia
11. Qatar
12. Oman (Muscat and Oman)
13. Yemen (Yemen Arab Republic)
14. Democratic Republic of Yemen (known previously as Aden, Federation of South Arabia, and Southern Yemen); includes Socotra Island
15. Bahrain
16. United Arab Emirates (formerly Trucial Sheikdoms or Trucial Coast)

Disputed Territory:

17. Kashmir (Jammu and Kashmir)—claimed by India and Pakistan

Neutral Zone:

18. Neutral Zone: Iraq/Saudi Arabia

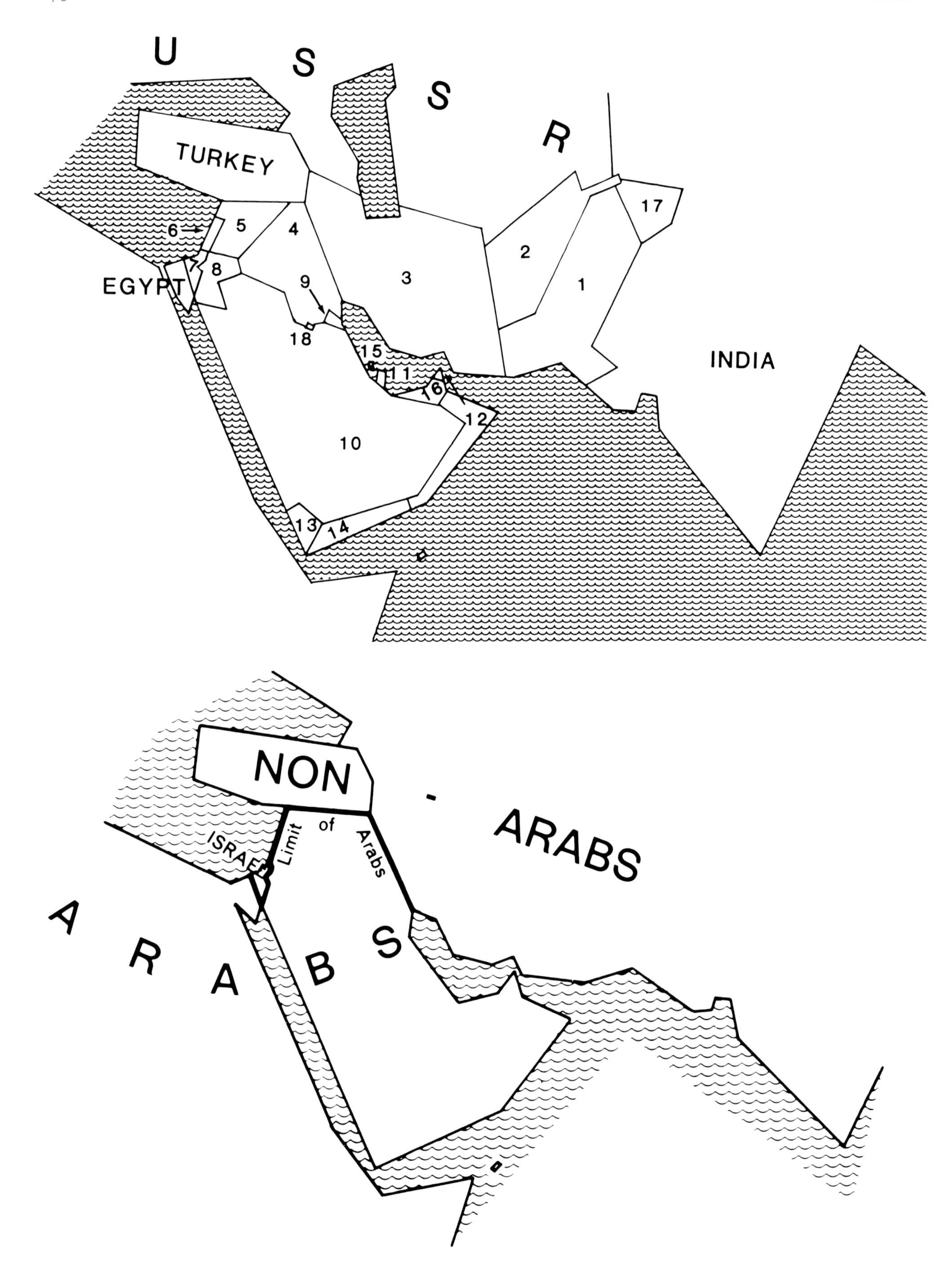

U S S R
TURKEY
17
6
5
4
2
EGYPT
7
8
3
1
9
18
INDIA
15
11
16
12
10
13
14
NON - ARABS
Limit of Arabs
ISRAEL
ARABS

Occupied Territories:

Israel occupies (as of this writing) territories in three neighboring states: *Golan Heights* (Syria), *Gaza Strip* (Egypt), and the *West Bank* (Jordan). See the map following. Label: Jerusalem, Tel-Aviv, and Haifa.

There is no "gimmick" for learning the political units of Southwest Asia. Careful study of the cartogram, as suggested, combined with an up-to-date atlas will be rewarding. Politically, *most* of the area is *unstable* and political affiliations change rather frequently. At this writing, Afghanistan is the only communist government, but it is under constant attack by anti-government rebels. Several states are openly pro-communist.

The major conflicts in the area of Southwest Asia today are:

a. *Iran:* social and governmental chaos; military power vacuum; oil reserves; ethnic uprisings (Kurds, Arabs, Azerbaydzhani); war with Iraq
b. *Afghanistan:* Soviet invasion (1980)
c. *Oil:* the Persian (Arabian) Gulf area is the world's primary supplier of petroleum; price increases imposed by the Organization of Petroleum Exporting Countries (OPEC) have been a key factor in world inflation

d. *Palestine:* annexation of East Jerusalem by Israel; Israeli settlements on the West Bank; the question of autonomy
e. *Lebanon:* the Christian-Muslim conflict
f. *Kashmir:* a Muslim land claimed by both India and Pakistan
g. *Human concerns:* a regional instability, caused by an appalling degree of illiteracy, hunger, poverty, disease, and staggering birth rates; many countries are not long emerged from feudalism

In countries where there is virtually *no* middle class, most political agitation stems from (1) the military, (2) the intellectuals, (3) the Muslim clergy, and (4) the students. The communists and extremist elements have no shortage of problems and issues to exploit, including western involvement in the area's oil resources, and western support of democratic (and highly European) Israel.

For years military stability in the area was more-or-less guaranteed by Israel, Turkey, and Iran, three strongly pro-western states. Iran has tumbled into virtual anarchy and Iraq (in 1980) reacted swiftly to the collapse of Iranian power. Turkey experienced a military coup in 1980. Only Israel (and possibly Egypt) seems capable of maintaining the power balance vis-a-vis the communists. Egypt may replace Iran in the balance-of-power equation in the region, but Iraq, despite its past record of leftist behavior, may receive enough support from more moderate Arab states to give it an edge over Egypt. The latter state has alienated itself from some of the Arab World because of its friendship with Israel.

The Middle East is a chessboard and one must anticipate the opponent's moves. A Russian move into Afghanistan, an Iraqi invasion of Iran, the seizing of American hostages, a maneuver by the American Navy in the Indian Ocean, a cut-off of oil supplies passing through the Strait of Hormuz—these are all moves in the "middle game." The objective of the geo-political "end-game" is "Checkmate," an English corruption of an old Persian (Iranian) term, "shāh māt," or "the Shah is dead."

The map on page 73 provides more detail than the cartograms you have studied. Refer to an atlas to assist you with the following.

Countries: Name every country on the map, including the partially mapped peripheral countries.

Water Bodies:

Gulf of Aden
Arabian Sea
Caspian Sea
Euphrates River
Mediterranean Sea
Nile River
Gulf of Oman
Persian (Arabian) Gulf
Red Sea
Shatt al'-Arab
Strait of Hormuz
Suez Canal
Tigris River

Cities: (capital cities are in *italics* and indicated by *stars* on the map; others are represented by *dots*)

Abādān, Iran
Abu Dhabi (Abu Zaby), United Arab Emirates
Aden (Madinat ash-Sha'b), Democratic Republic of Yemen
Ahvāz, Iran
Ammān, Jordan
Baghdad, Iraq
Basra (Al Basrah), Iraq
Beirut (Bayrūt), Lebanon
Cairo (Al Qāhirah), Egypt
Damascus (Dimashq), Syria
Doha (Ad Dawhah), Qatar
Jerusalem (Yerushalayim) Israel
Khorramshahr, Iran
Kuwait (Al Kuwayt), Kuwait

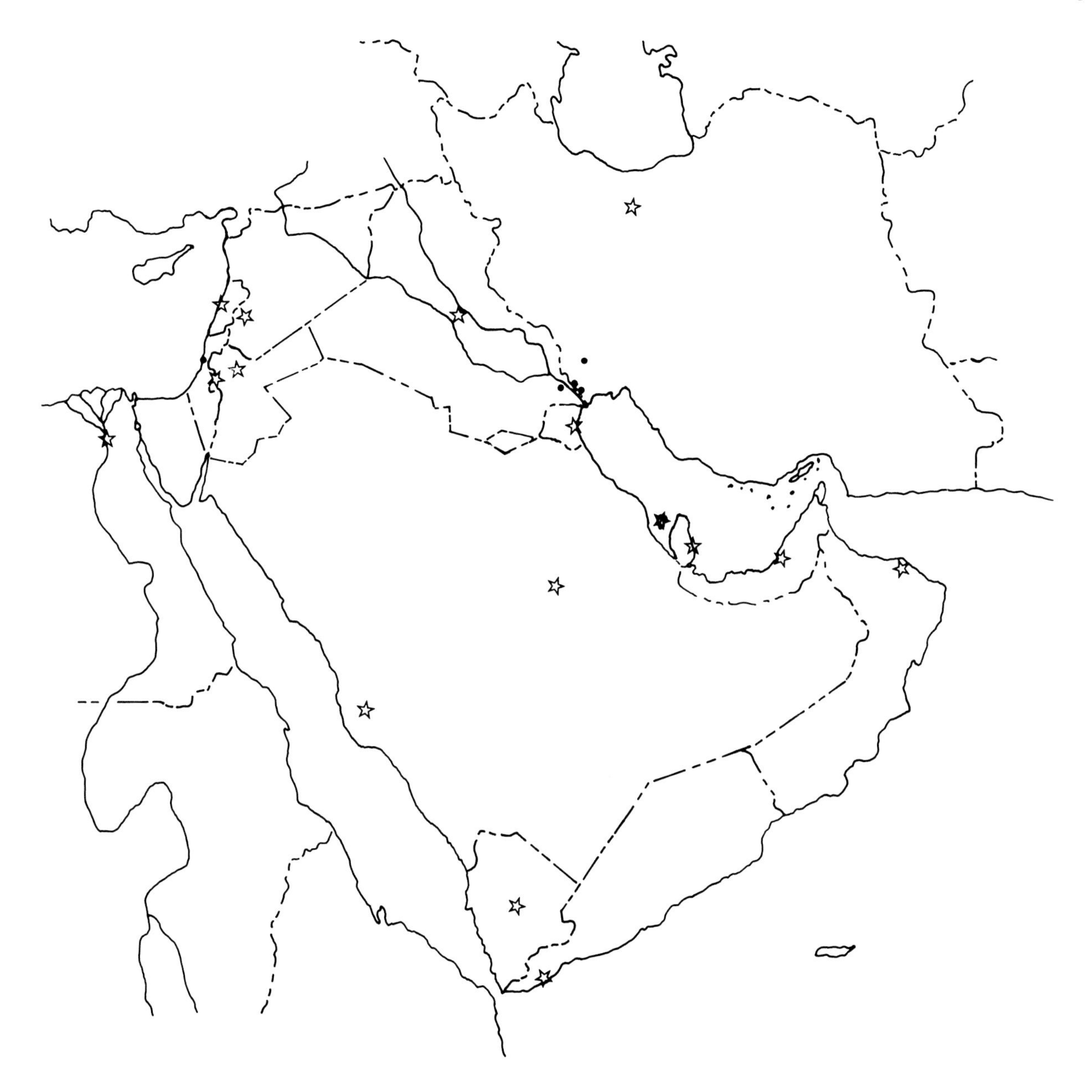

Manama (Al Manamah), Bahrain
Mecca (Makkah), Saudi Arabia (Saudi Arabia has two capitals)
Muscat (Masqat), Oman
Riyadh (Ar Riyād), Saudi Arabia (Saudi Arabia has two capitals)
San'a, Yemen Arab Republic
Tehrān, Iran
Tel Aviv-Jaffa (Tel Aviv-Yafo), Israel

A Suggestion:

Locate the principal oil fields; locate the ethnic Arab population of Iran; locate the area occupied by the Kurds in Iran-Iraq-Turkey; locate the Sinai Peninsula; study a good map of Afghanistan and ponder the questions raised by Soviet domination of this state.

Self-Test

(Answer yes or no.)

__________ 1. Do you know and can you locate the 17 political units of Southwest Asia? (Disputed territory is not included).

__________ 2. Do you know and can you locate the major water bodies and cities given above?

__________ 3. Do you know the reasons why Southwest Asia is an area of continuing crises?

__________ 4. Can you distinguish among the following and explain each?

____________ a. Middle East

____________ b. Near East

____________ c. Arab World

____________ d. non-Arab portion of Southwest Asia

____________ e. Muslim portion of Southwest Asia

If unable to answer *yes* to all of these questions, this section should be reviewed *thoroughly* before continuing. Although Southwest Asia is a complex realm, the one to follow (Southern and Southeast Asia) may well be more complex. Since this is the case, it would be unwise to move on until this portion of the text has been mastered.

II. Southern and Southeast Asia

Sixteen political units form the realm of *Southern and Southeast Asia.* Once again, this is an extremely complex political area. Most of these states were under European colonial administration until after World War II. As with Southwest Asia, this area is "emerging," and is subject to all of the strains and stresses that new nations always experience, *plus* the ever-present East-West conflict.

Refer to the cartogram that follows while studying the list below.

1. India (1a. Kashmir: disputed between India and Pakistan)
2. Nepal
3. Bhutan
4. Bangladesh
5. Republic of Maldives
6. Sri Lanka (formerly Ceylon)
7. Burma
8. Thailand (the *only* country in this area *never* under European control)
9. Laos
10. Kampuchea (Cambodia)
11. Vietnam
12. Philippines
13. Malaysia (Malaya, Sarawak, Sabah)
14. Singapore
15. Indonesia (includes over 3,000 islands)
16. Brunei

Papua New Guinea (17) is shown for comparative purposes. It is discussed in chapter 9.

Note: The tiny Himalayan kingdom of *Sikkim* (between Nepal and Bhutan) was absorbed by India in 1974. The Portuguese colony of *Timor* (on the island of Timor) was annexed by Indonesia in 1976.

Each student is urged to study the maps of Southern and Southeast Asia in the atlas *very carefully.*

Locate the following on the cartogram. Draw in the rivers and use dots for city-symbols.

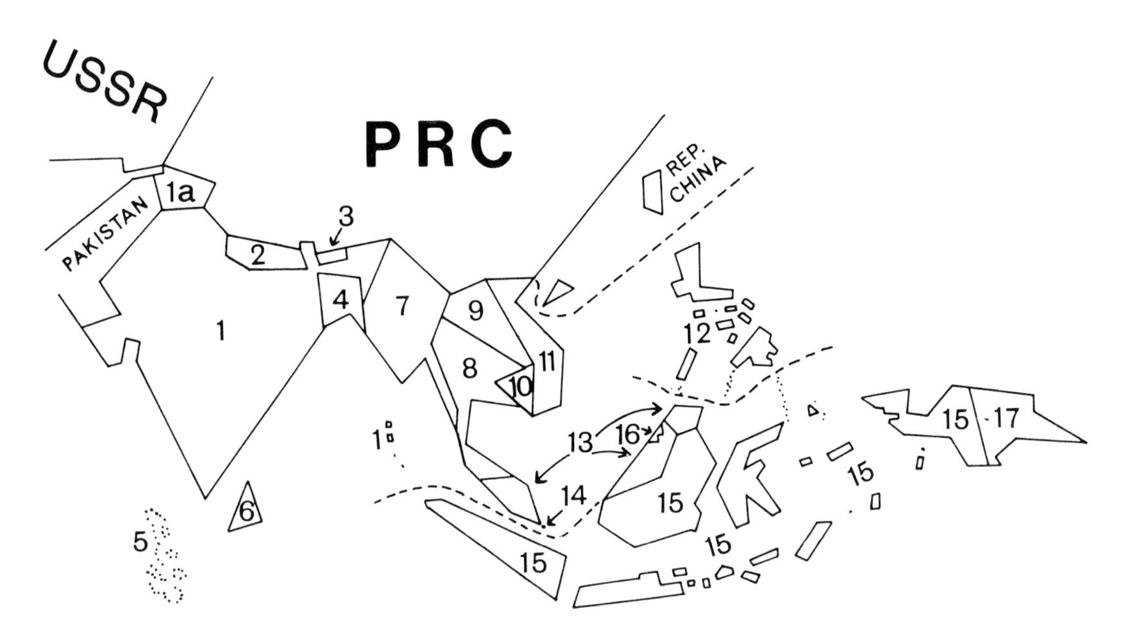

Water Bodies:

- Indian Ocean
- Bay of Bengal
- Gulf of Siam
- South China Sea
- Gulf of Tonkin
- Ganges River
- Irrawaddy River
- Menam (Chao Phraya) River
- Mekong River
- Red River (Song Hong)

Terrain Features:

- Himalaya Mountains
- Ganges Basin
- Ganges Delta
- Deccan Plateau
- Thar Desert (Great Indian Desert)

Islands:

- Sumatra, Indonesia
- Java, Indonesia
- Borneo (Kalimantan), Indonesia/Malaysia
- Celebes (Sulawesi), Indonesia
- New Guinea, Indonesia/Papua New Guinea
- Hainan, People's Republic of China
- Taiwan (Formosa), Republic of China
- Luzon, Philippines
- Mindanao, Philippines

Cities: (Capitals in italics)

- Karachi, Pakistan
- *Islamabad,* Pakistan
- *New Delhi,* India
- Calcutta, India
- *Dacca,* Bangladesh
- *Rangoon,* Burma
- *Bangkok (Krung Thep),* Thailand
- *Vientiane,* Laos
- *Phnom Penh,* Cambodia (Kampuchea)
- *Hanoi (Ha Noi),* Vietnam
- Ho Chi Minh City (Saigon), Vietnam
- *Colombo,* Sri Lanka
- *Kuala Lumpur,* Malaysia
- *Singapore,* Singapore
- *Djakarta,* Indonesia
- *Manila,* Philippines

The significance of the rivers listed and located cannot be overemphasized. The flood plains and deltas of these (and *many* more!) rivers of Southern and Southeast Asia produce the *rice* that "powers" the realm. There is a *direct* relationship between *rivers* and *population density.* Southeast Asia (Burma to Vietnam) is the *only area* in Asia producing a *surplus* of rice.

This is not only an agricultural area. The *Malayan Peninsula* is the world's largest supplier of *tin* and *rubber,* and major petroleum deposits are located in *Sumatra* (Indonesia) and *Brunei.*

No brief self-test could be devised that will convey the real importance of this part of the world. Review this section *completely* and *thoroughly* before continuing. Supplement this review with a *close* examination of a good, detailed map of the area.

III. East Asia

At least one-fourth of *all* the people on earth live in the realm called *East Asia.* This is an area dominated by the People's Republic of China and Japan.

The cartogram below reveals that East Asia is, politically at least, much more simple than the two other Asian realms that have already been examined. The political breakdown is as follows:

3 sovereign communist states:

People's Republic of China (usually abbreviated *PRC*)
People's Republic of Mongolia (PRM)
People's Democratic Republic of Korea (North Korea)

3 sovereign non-communist states:

Japan
Republic of China (on the island of Taiwan, or Formosa)
Republic of Korea (South Korea)

2 colonial territories on the Chinese mainland:

Hong Kong (a Crown Colony of the United Kingdom)
Macau (Macao, an Overseas Province of Portugal)

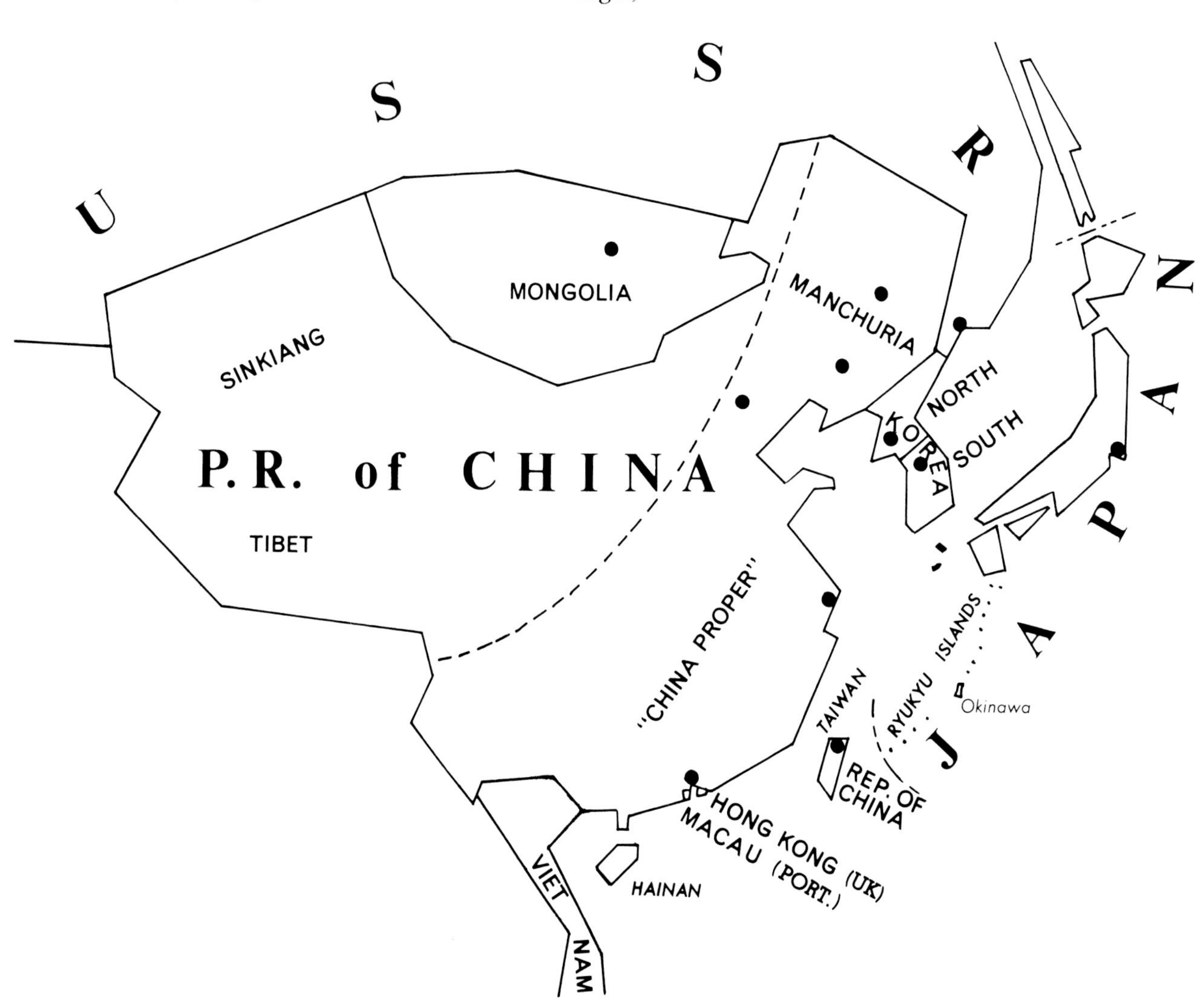

The *People's Republic of China (PRC)* is the most populous nation on earth. *One out of every five* human beings lives in the PRC (total population: approximately 1,000,000,000). In area, the PRC ranks *third,* behind the USSR and Canada. With an area of 9,699,567 km² (3,746,453 mi²), the PRC is slightly smaller than Canada and slightly larger than the United States.

The territory comprising the PRC today incorporates a large area that is not (and has never been) culturally Chinese. Historically, Chinese culture developed in the eastern part of the country and generally south of Beijing (Peking). This region is often called China Proper (refer to the cartogram on page 76).

The area known as *Manchuria* has become Chinese only in the past few centuries, as Chinese migrated northward and gradually replaced the Manchus. At one time in the past the Manchus were a distinctive culture, speaking a language related to central Asian tongues and completely different from Chinese. Today, Chinese culture is dominant *east* of the dashed line shown on the cartogram. Three provinces have replaced the former Manchuria.

The line referred to divides the PRC into a *humid east* and an *arid west.* Inasmuch as the Chinese have long been agricultural peoples, there was a tendency to remain in the humid east. West of this line the population was once largely pastoral and nomadic, and these activities are still characteristic of large areas.

The people of *Xinjiang (Sinkiang)* are closely related to the people across the border in the USSR. They speak similar languages (non-Chinese) and share the *Muslim* religion with their cousins in Soviet Asia. This zone of Russian-Chinese conflict (cultural as well as political) has already been mentioned (page 62).

Mongolia, claimed by China for centuries, became a sovereign state this century, though some Mongolians still reside within the PRC in an area called *Inner Mongolia (Neimenggu).* Mongolia has become "Russianized" to such an extent that it is virtually a satellite state of the USSR. As in Sinkiang, the people of Mongolia speak non-Chinese languages. However, they are not Muslims, but practice their own version of *Buddhism.*

Mongolians and Manchus were traditional enemies of the Chinese in past times, and the *Great Wall* of China was possibly built to keep pastoral Manchus and Mongolians from invading agricultural China Proper.

Xizang (Tibet), long a remote religious state on the world's highest plateau, has only recently been fully incorporated by the PRC. As with the other outlying peoples of the PRC, the Tibetans have little in common with their Chinese master. Tibetan *Buddhism* is unique, and although the language is related to Chinese, it is so far removed that Tibetans and Chinese cannot converse.

Even Chinese speakers have difficulty communicating with the spoken word, for there are many "Chinese" languages. They do, however, have a common system of ideographs, which gives unity to the written Chinese language. Non-Chinese in the western part of the country do not share either verbal or written languages with the Chinese.

The *Republic of China* occupies the island of *Taiwan (Formosa).* It is a sovereign state and still, technically, in a state of war with the PRC. Though ousted from the United Nations (and replaced by the PRC in that body), the Republic of China exists as a politically independent, viable state. It is an important industrial nation and enjoys one of the highest standards of living in East Asia.

The *People's Democratic Republic of Korea (North Korea)* and the *Republic of Korea (South Korea)* are products of World War II, where a homogeneous people were divided by the East-West conflict. South Korea has followed a pattern of development similar to that in Japan and the Republic of China; North Korea has evolved along typical communist lines. There is always a temptation to make a geographical comparison of Korea with Vietnam. Eventually, some sort of accord will be worked out between the two Koreas. The question is, will it be a peaceful one? One could add, what sort of accord will be reached between the two Chinas?

Hong Kong and *Macau (Macao)* are relics of European colonialism. Hong Kong is a British colony, acquired in 1842. Hong Kong is an island, upon which is situated the colonial capital of *Victoria. Kowloon,* a part of the colony, is a peninsula of the mainland, just north of the island. Additional territory *(New Territories)* was leased from China in 1898. The colony will return to the PRC in 1999.

Macau is a Portuguese possession, the only overseas territory still occupied by that once-powerful colonial state. It lies across the estuary from Hong Kong and has been Portuguese since 1557. In 1999 control reverts to the PRC.

Study the cartogram on page 76. *Add to it the following:*

Cities: indicated by dots on the cartogram; capitals in italics)

Ulan Bator (Ulaanbaatar), PR of Mongolia
Haerbin (Harbin), PRC
Shenyang (Mukden), PRC
Beijing (Peking), PRC
Shanghai, PRC
Guangzhou (Canton), PRC
T'aipei, R. of China
P'yongyang, North Korea
Seoul (Sŏul), South Korea
Vladivostok, USSR
Tokyo, Japan

Rivers: (Use the atlas.)

Amur
Yalu
Hwang (Huang)
Yangtze (Chang)
Si (Xi, Hsi, Hsün)

Water Bodies:

Sea of Japan
Yellow Sea
East China Sea
South China Sea
Formosa Strait
Gulf of Tonkin

Mountains:

Himalayas
Tien Shan (Tyan Shan)
Qinlingshan (Tsingling, Ch'in Ling)
Altaj (Altai)

Gobi (Desert)

Japan merits a special note, for it is the closest economic rival that the United States has today. It leads the world in shipbuilding, optical goods, radio and television receivers, and probably automobile production; it is second in synthetic rubber, chemical textiles, aluminum, and steel production. And the list could be continued for several pages. This is a truly remarkable achievement when one recalls that Japan's economy was in ruin in 1945 and the nation had only 17 ocean-going ships still afloat.

Refer to the following map as you follow the discussion below. You should also have an atlas and may find Map 8, *East Asia* (at the end of the book) helpful.

Japan lies east of the Asian mainland and consists of four main islands, numerous small ones, and the important *Ryukyu Islands* that extend southwestward. *Label the map* to indicate the main Japanese islands of *Hokkaido* (northernmost), *Honshu* (largest), *Shikoku* (smallest), and *Kyushu* (southernmost).

The water body between Honshu and Shikoku is the *Inland Sea. Label this on the map.* The area surrounding the Inland Sea is the Japanese *culture hearth.* This is where Japanese culture was "born," as oceanic culture, moving northward through the Ryukyu Islands, blended with mainland culture coming through Korea from China more than 2,000 years ago. The first permanent capital *(Nara)* was located near the eastern end of the Inland Sea, on the island of Honshu and just east of the modern city of *Osaka.* This was in the early eighth century A.D. In A.D. 794 a new capital was built at *Kyoto,* north of Nara. It remained the capital for 700 years.

Locate the Nara-Osaka-Kyoto area on the map. This is one of the four major industrial districts in modern Japan (the *Kinki District).* The city of *Kobe* is also in this district. The other three primary industrial districts are: the northern tip of Kyushu, the *Tokyo* area, and the *Nagoya* district (located between Tokyo and Osaka). Over 60 percent of all Japanese live in these four districts. *Locate these three industrial districts on the map.*

By the end of the sixteenth century a new capital was established at Edo (now Tokyo). Europeans (Portuguese) arrived in 1543 and, although Japan barred Europeans from 1639–1853, the floodgates were opened to European technology after 1868. Japan had about 40,000,000 people then. A lowered death rate (due to western medical technology) caused the population to climb to 60,000,000 by the 1920s. Since 1925 it has increased by almost 1,000,000 a year.

Japan now ranks *seventh* in population among world nations. Though smaller in area than California, its population is *four times* as large as that of our most populous state! In terms of population *density,* Japan has 330 people per square km (855/mi²). Although this is a high figure, it is less than some European nations. But in terms of people per unit of cultivated land, Japan's density reaches 2,000 people per square km (5,200/mi²).

When Japan began to industrialize a little over a century ago it had sufficient food and mineral resources. As population grew and resources diminished Japan turned to colonialism to solve its dilemma. After the collapse of the empire (1945), the nation turned to foreign trade, mainly with the United States and other western countries. By the 1980s it was importing about 25 percent of its food, over half of its coal, and almost all of its iron ore, bauxite, and petroleum.

In many ways, Japan may be compared with the United Kingdom. Both states are island nations; both are near densely populated continents; both have large populations; both are industrial; both have lost the colonial sources of raw materials; both must import and export; both have had to reevaluate their relationships with their neighboring continents. Can you think of ways the two great states differ?

7

Africa

No other continent has witnessed so many name changes during the last two decades as has Africa. These have occurred as a direct result of sweeping political alterations that began after World War II. Before that time there were only *four* independent countries on the continent (Liberia, Ethiopia, Egypt, and the Union [now Republic] of South Africa). The remainder of the continent was colonial territory, shared by the United Kingdom, France, Italy, Portugal, Spain, and Belgium. (Germany had held colonies but lost them during World War I.) Today, there are *no* colonial territories in Africa.

With the help of your atlas, locate on the cartogram (page 82) each of the countries listed. *Caution:* Watch out for old names and alternate spellings in your atlas!

POLITICAL UNITS OF AFRICA AND THEIR CAPITAL CITIES

North Africa:

Morocco .... *Rabat*
Algeria .... *Algiers (Alger)*
Tunisia .... *Tunis*
Libya .... *Tripoli (Tarābulus)*
Egypt .... *Cairo (Al Qāhirah)*

Second Tier:

Mauritania .... *Nouakchott*
Mali .... *Bamako*
Burkina Faso .... *Ouagadougou*
Niger .... *Niamey*
Chad .... *N'Djamena*
Central African Republic .... *Bangui*
Sudan .... *Khartoum (Al Khurtūm)*
Ethiopia .... *Addis Ababa (Addis Abeba)*
Djibouti .... *Djibouti*
Somalia .... *Mogadisho*

West Coastal:

Senegal .... *Dakar*
Gambia .... *Banjul*
Guinea-Bissau .... *Bissau*
Guinea .... *Conakry*
Sierra Leone .... *Freetown*
Liberia .... *Monrovia*

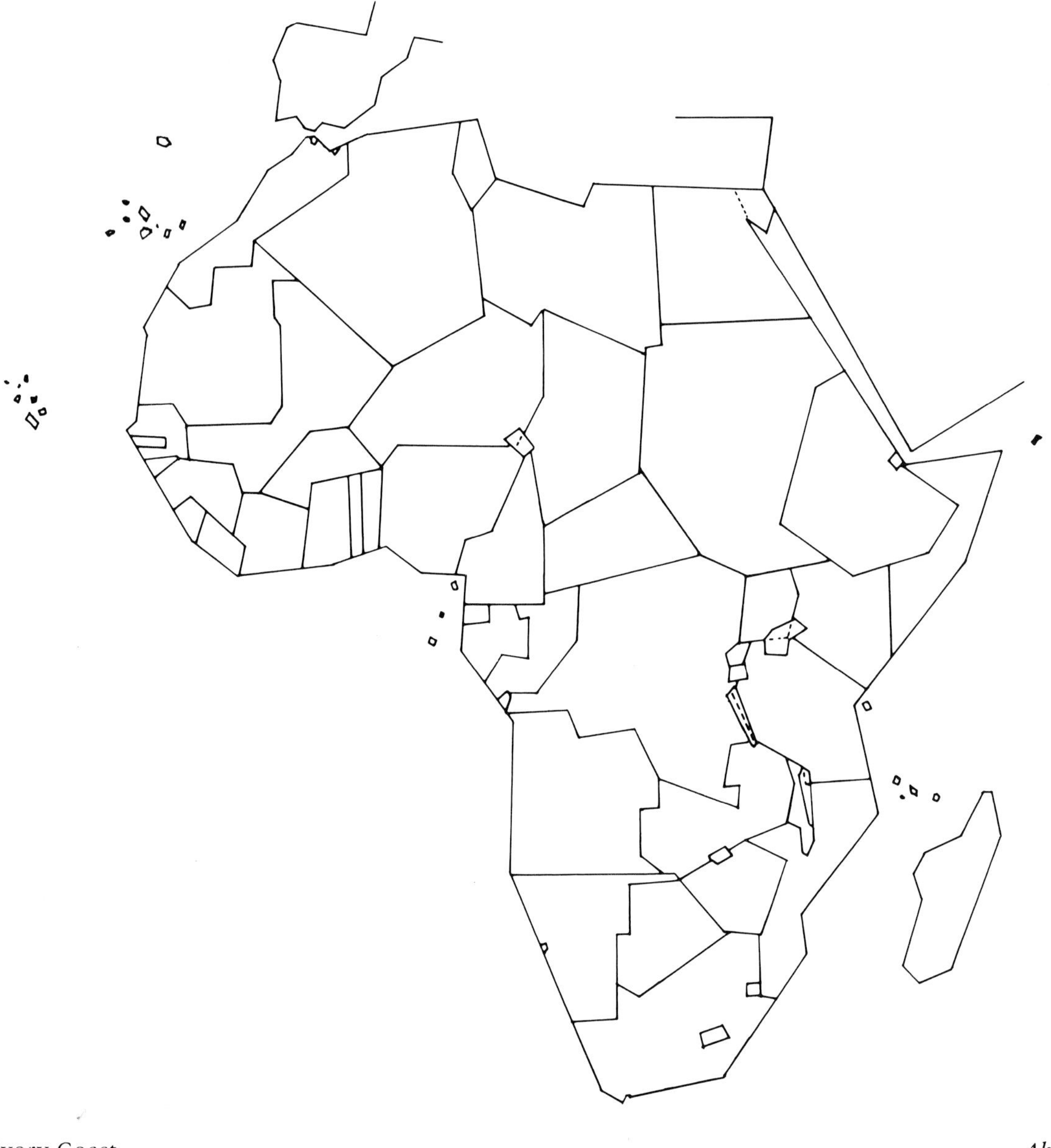

Ivory Coast .... *Abidjan*
Ghana .... *Accra*
Togo .... *Lomé*
Benin (formerly Dahomey) .... *Porto-Novo*
Nigeria .... *Lagos*
Cameroun (Cameroon) .... *Yaoundé*
Equatorial Guinea .... *Malabo*
Gabon .... *Libreville*
Congo .... *Brazzaville*
Zaïre .... *Kinshasa*
Angola (inc. Cabinda) .... *Luanda*

East Africa:

Kenya .... *Nairobi*
Tanzania .... *Dodoma*
Uganda .... *Kampala*

Rwanda .... *Kigali*
Burundi .... *Bujumbura*
Zambia .... *Lusaka*
Malawi .... *Lilongwe*
Mozambique .... *Maputo*

South Africa:

Zimbabwe .... *Harare*
Botswana .... *Gaberone*
Swaziland .... *Mbabane*
Lesotho .... *Maseru*
Republic of South Africa .... *Pretoria* and *Cape Town*

Islands (counterclockwise from the northwest):

Madeira Islands; NOT African; integral part of Portugal
Canary Islands; NOT African; integral part of Spain
Cape Verde Islands .... *Praia*
São Tomé and Príncipe .... *São Tomé*
Madagascar .... *Antananarivo*
Comoros .... *Moroni*

Special Situations:

Ceuta and Melilla: Cities that are part of metropolitan Spain; enclaves on Moroccan coast; NOT colonial territories.

Namibia (South West Africa). Presently under the administration of the Republic of South Africa; capital is *Windhoek.*

Walvis Bay: Enclave on coast of Namibia; part of the Republic of South Africa; will probably NOT be included in Namibia if the latter achieves independence.

Comoros: Comoros became independent in 1975, but one of the four islands, Mayotte, voted to remain under French administration. It is an Overseas Department.

Africa's political units are difficult to learn—there is no argument on that score. It is probably not necessary to learn every capital, but everyone should know *all* of the countries *and the items listed below.* Careful study with an up-to-date atlas will pay great dividends. Your instructor may wish to augment the list that follows. Locate these on the *map* of Africa (page 84).

Lakes:

Victoria
Tanganyika
Malawi (Nyasa)
Kariba (man-made; behind Kariba Dam)
Chad

Rivers:

Nile
Congo (Zaïre)
Limpopo
Orange
Zambezi
Senegal
Niger

Other Water Features:

Aswan Dam
Suez Canal
Victoria Falls
Gulf of Guinea
Mediterranean Sea
Red Sea
Gulf of Aden
Mozambique Channel

Mountains:

Atlas
Drakensberg
Mt. Kilimanjaro (5,895 m [19,340 ft.])

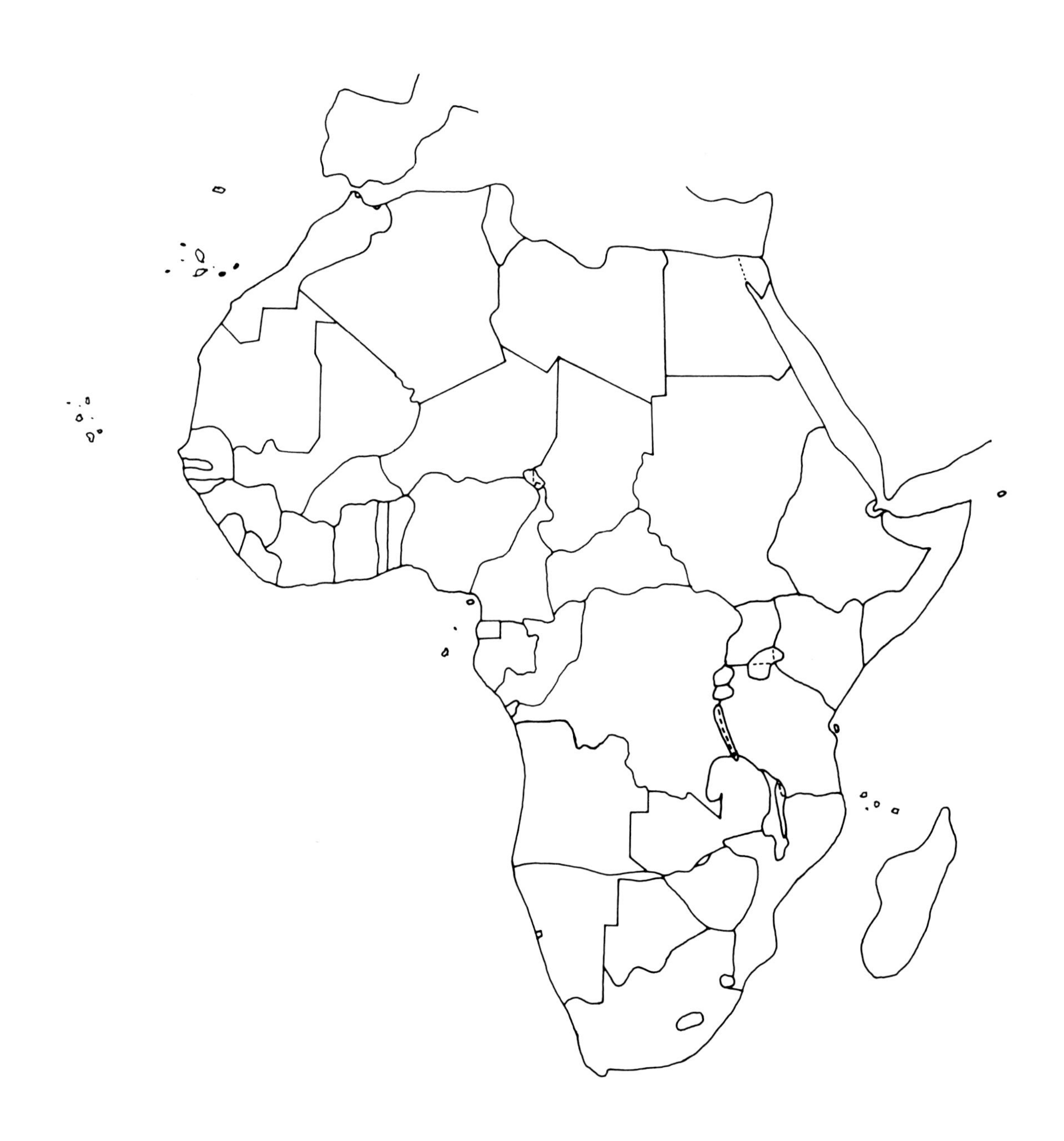

Subregions:

Sahara
Congo (Zaïre) Basin
Katanga (Shaba)
Kalahari
Guinea Coast

Selected Cities (capitals in italics):

Cairo (Al Qāhirah)
Alexandria (Al Iskandarīyah)
Tripoli (Tarābulus)
Tunis
Algiers (Alger)
Lagos
Addis Ababa (Addis Abeba)
Nairobi
Mombasa
Kinshasa
Harare
Pretoria
Cape Town
Johannesburg

Politics and *economics* are often inseparable, and the more important African places that have been cited are focal points of both. *Racial unrest* in the southern part of the continent is tied to each, and control of one of these spheres of activity usually means control of the other. Manipulations of the world's petroleum supply by the Organization of Petroleum Exporting Countries (OPEC) is another example of political and economic merger. *Three of the top ten oil producers are in Africa* (Libya, Algeria, and Nigeria). Libya, for example, has used its oil profits to export its own version of extremist political behavior.

The Republic of South Africa, while lacking in oil, is richly endowed with many other resources. It leads the world in *gold* and *diamond* production, has abundant *coal,* and is the most developed nation on the continent. In a country where there are more Europeans than anywhere else in Africa, but where they are in a minority by four or five to one, the stakes can be very high.

From the north (where the Egyptians have been a force for peace and progress, and the Libyans have been supportive of terrorism and aggression), to the east (Ethiopia) and west (Angola) where Cuban troops have enforced a new form of communist imperialism, to the south (where Black Africa and White Africa have squared off), the continent is in a state of flux. The potential for tremendous development is present, but without political and social stability there can be little progress. To a degree this has been achieved along the Guinea Coast in places (such as Ivory Coast), and in East Africa (especially in Kenya), but the emergence of many states has barely begun.

Self-Test

1. Can you locate the countries of Africa on an outline map?
2. What are the three major oil-producing states in Africa?

_______________, _______________, _______________

3. Where is racial unrest at its greatest in Africa today? ______________________

   Why? ______________________

4. Name seven large African rivers. ____________, ____________, ____________, ____________, ____________, ____________, ____________

5. What is the capital of each of the following countries?

   Republic of South Africa (2 capitals): ____________, ____________

   Egypt: ____________

   Zimbabwe: ____________

   Nigeria: ____________

   Kenya: ____________

   Zaïre: ____________

6. What does OPEC stand for? ______________________

   ____

7. How many European colonial possessions are now on the African mainland? ____________

8. What is the status of Ceuta and Melilla? ______________________

   ____

9. What is another name for Namibia? ____________

10. What do Cape Verde, São Tomé and Príncipe, Comoros, and Madagascar have in common? ____________

    ____

11. What is the most developed country in Africa? ______________________

12. Where is the largest population of European extraction in Africa? ______________________

8

Australia and New Zealand

Australia and *New Zealand,* most prosperous and highly developed nations of the Southern Hemisphere, lie half a world away from their mother country, England. Today, their roles in world affairs are becoming more sharply defined, for these nations front on one of the world's most critical doorsteps—Southeast Asia.

Australia is the sixth largest country on earth in area (about the same size as the 48 contiguous United States). With a population of only 16,000,000, however, it is considered to be an underpopulated nation. This is significant, for just to the north 150,000,000 Indonesians are jammed into much less territory, and the PRC hovers over the entire area with an estimated 1,000,000,000 people. At the present state of technology, not all of Australia can hold people though, for some 80 percent of the land is *arid.* Australia's people are concentrated in the humid east, southeast, and southwest.

New Zealand lies 1,900 km (1,200 mi) southeast of Australia. It is comprised of several islands, the largest being *North Island* and *South Island.* Each of the two larger islands is about 800 km (500 mi) long. The country supports a population of 3,300,000 at one of the world's highest living standards.

Australia is probably more like the United States than any other foreign area. It is about the same size, has similar traditions, and is organized politically (states and a capital district) like the United States. Also, Australia is an urban nation with an economy well balanced between agriculture, livestock, and growing industrialization.

Locate the following on the outline map on page 89. Use the atlas as necessary.

AUSTRALIA:

States and their Capitals:

Western Australia .... *Perth*
South Australia .... *Adelaide*
Queensland .... *Brisbane*
New South Wales .... *Sydney*
Victoria .... *Melbourne*
Tasmania .... *Hobart*

Territory:

Northern Territory .... *Darwin*
Australia's major interior city .... Alice Springs

Australian Capital Territory:

Canberra (this is a federal district, like the District of Columbia, and is not in any state).

Jervis Bay, about 160 km (100 mi) south of Sydney, is also a part of the A.C.T. The Royal Australian Naval College is located there.

Water Bodies:

Indian Ocean
Tasman Sea
Coral Sea
Gulf of Carpentaria
Great Australian Bight
Bass Strait
Murray-Darling River

Physiographic Features:

Great Barrier Reef
Great Artesian Basin
Great Dividing Range
Australian Alps

NEW ZEALAND:

Cities:

Wellington (capital)
Auckland (largest city)
Christchurch

Other Features:

North Island
Cook Strait
South Island
Southern Alps

Adjacent Islands:

Indonesia

New Caledonia (France)

Loyalty Islands (France)

Vanuatu (formerly New Hebrides)

Solomon Islands (except Bougainville)

Papua New Guinea (eastern half of the island of New Guinea, plus the Admiralty Islands, New Ireland, New Britain, and Bougainville)

Self-Test

1. Name the states of Australia: ____________, ____________, ____________, ____________, ____________, ____________.

2. What is the large territory in Australia called? ____________

3. A.C.T. is an abbreviation for what? ____________

4. New Zealand consists of two main islands called: ____________ and ____________

5. The capital of New Zealand is: ____________

6. The capital of Australia is: ____________

7. The sea between Australia and New Zealand is the ____________ Sea.

8. The principal river system of Australia is called the: ____________

9

Islands and Oceans

I. Pacific

Study the Pacific Ocean cartogram below.

Nesia means islands.
Poly means many.
Micro means small.
Mela means black; i.e., inhabited by dark-skinned Australoids. Review carefully, then take the self-test on the next page.

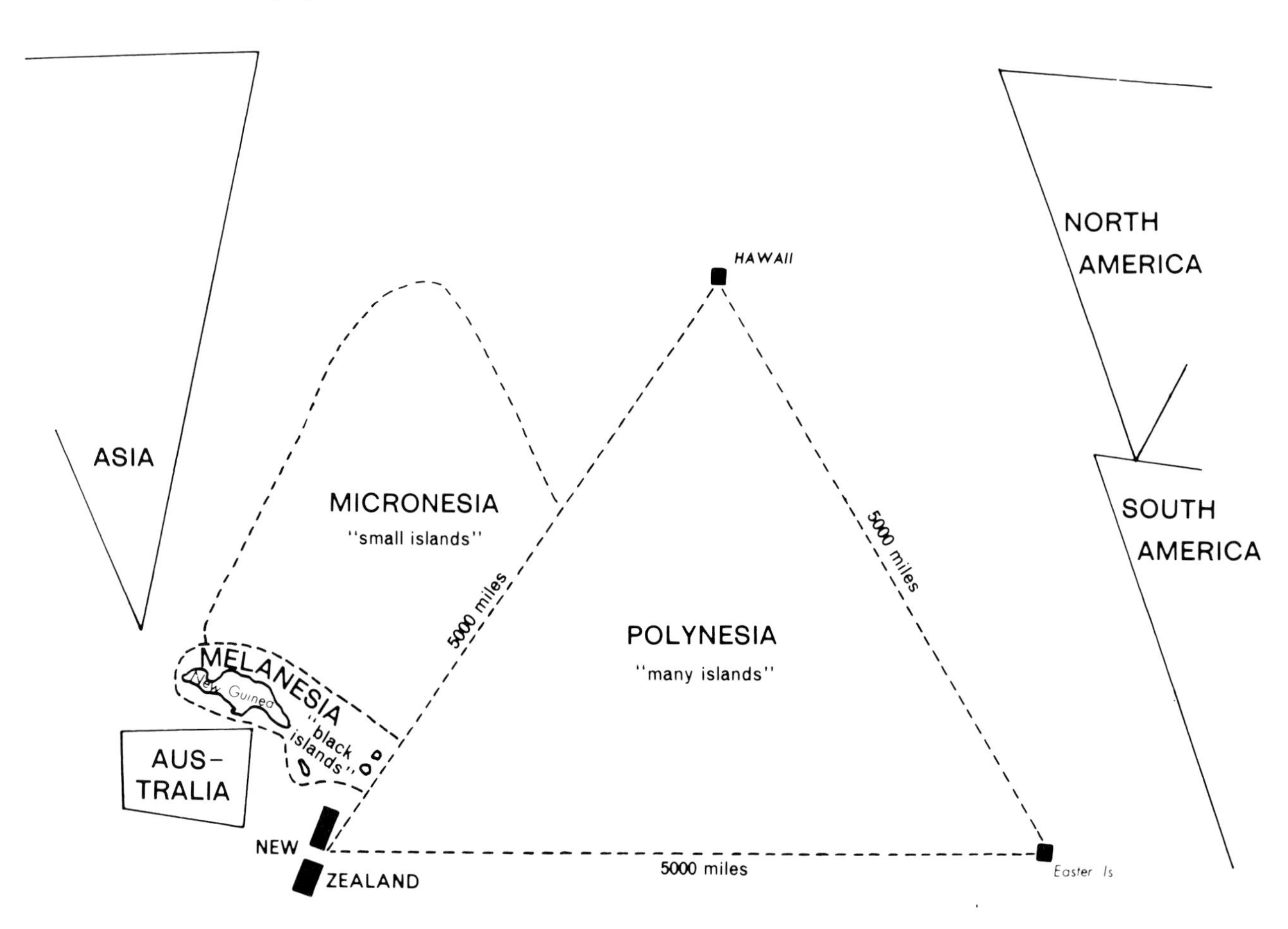

Self-Test

1. Polynesia forms a vast triangle defined by the ________________ Islands in the north; ____________ Island on the southeast; and ________________ on the southwest.
2. *Nesia* means ________________
3. The Pacific region of *many islands* is called ________________
4. The Pacific region of very *tiny* islands is called ________________
5. The Pacific islands inhabited by Australoids ("blacks") are called ________________ ________________.
6. The island region just north of Australia is ________________
   Check these answers with the cartogram on page 90.

Study pages 91–92 in conjunction with a good map of the Pacific area.

There are *nine* sovereign states in the Pacific Ocean realm (*excluding* the large fringing island countries of New Zealand, Philippines, Japan, Indonesia, and Taiwan).

The Nine Independent States and Their Capitals:

Fiji .... *Suva*
Kiribati (formerly Gilbert Islands) .... *Bairiki*
Nauru .... *Domaneab*
Papua New Guinea .... *Port Moresby*
Samoa (formerly Western Samoa) .... *Apia*
Solomon Islands .... *Honiara*
Tonga .... *Nuku'alofa*
Tuvalu (formerly Ellice Islands) .... *Funafuti*
Vanuatu (formerly New Hebrides) .... *Port-Vila*

All of the other Pacific islands are either colonies, protectorates, dependencies, or self-governing territories administered by one of the countries listed below.

Australia	Mexico
Chile	New Zealand
Ecuador	USSR
France	United Kingdom
Japan	United States

Not all of the islands are indicated on p. 93. The *Mexican* islands lie just off the Mexican coast and are not mapped. Some of *Chile's* possessions are also unmapped. The Galápagos Islands, belonging to Ecuador, are shown on p. 37. The Soviet-controlled Kurile Islands extend northward from Japan.

Not all Pacific islands are microscopic. *New Zealand* is as large as Colorado, and *New Guinea* is the second largest island in the world (Greenland is larger). *Fiji* and *New Caledonia* are each larger than Connecticut.

A NOTE ABOUT THE ISLANDS AFFILIATED WITH THE UNITED STATES

Over 2,000 Pacific islands, ranging in area from less than a square km^2 ($0.4\ mi^2$) to more than 500 km^2 (200 mi^2) are affiliated with the United States in one way or another. The vast majority of these are uninhabited coral islands.

The U.S. Pacific islands may be classified as follows:

Island	*Number of Islands*	*Capital*
Commonwealth of the Northern Mariana Islands	21	*Saipan*
Republic of the Marshall Islands (Freely Associated State)	1,225	*Majuro*
Federated States of Micronesia (Freely Associated State)	607	*Pohnpei*
Republic of Palau (Strategic Trusteeship; will become Freely Associated State)	350	*Koror*
American Samoa (Unincorporated Territory)	7	*Pago Pago*
Guam (Unincorporated Territory)	1	*Agana*
Small Populated Islands		
Wake, Wilkes, and Peale (USAF administration)	3	
Midway (USN administration)	2	
Johnston Atoll (USN administration)	1	
Small Uninhabited Islands		
Kingman Reef (USN administration)	1	
Howland, Jarvis, Baker, Palmyra (Interior Dept. administration)	4	

The Northern Marianas, Marshalls, Federated States, Palau, Samoa, and Guam are all self-governing. Residents of the Northern Marianas and Guam are U.S. citizens; Samoans are U.S. nationals.

The Northern Marianas, Guam, and Samoa each send one delegate to the U.S. House of Representatives.

The Freely Associated States (including Palau) are self-governing, sovereign states in every way except for defense. The U.S. assumes this responsibility.

Most of the residents of the small populated islands are U.S. citizens on temporary assignment (military, Atomic Energy Commission, weather personnel, airline employees, etc.).

Data for the major populated islands and island groups are given in the Appendix.

II. Atlantic

The sovereign island-states in the Atlantic Ocean have already been mentioned (*Bahamas,* pp. 31–32; *Cape Verde Islands,* p. 83; *Sao Tome and Principe,* p. 83; and part of *Equatorial Guinea,* p. 82; the islands of *Masie Nguema Biyogo* [formerly Fernando Póo] and *Pagalu* [formerly Annobón] are parts of Equatorial Guinea). Excluded, of course, are the large, traditional oceanic states, such as Iceland and the United Kingdom.

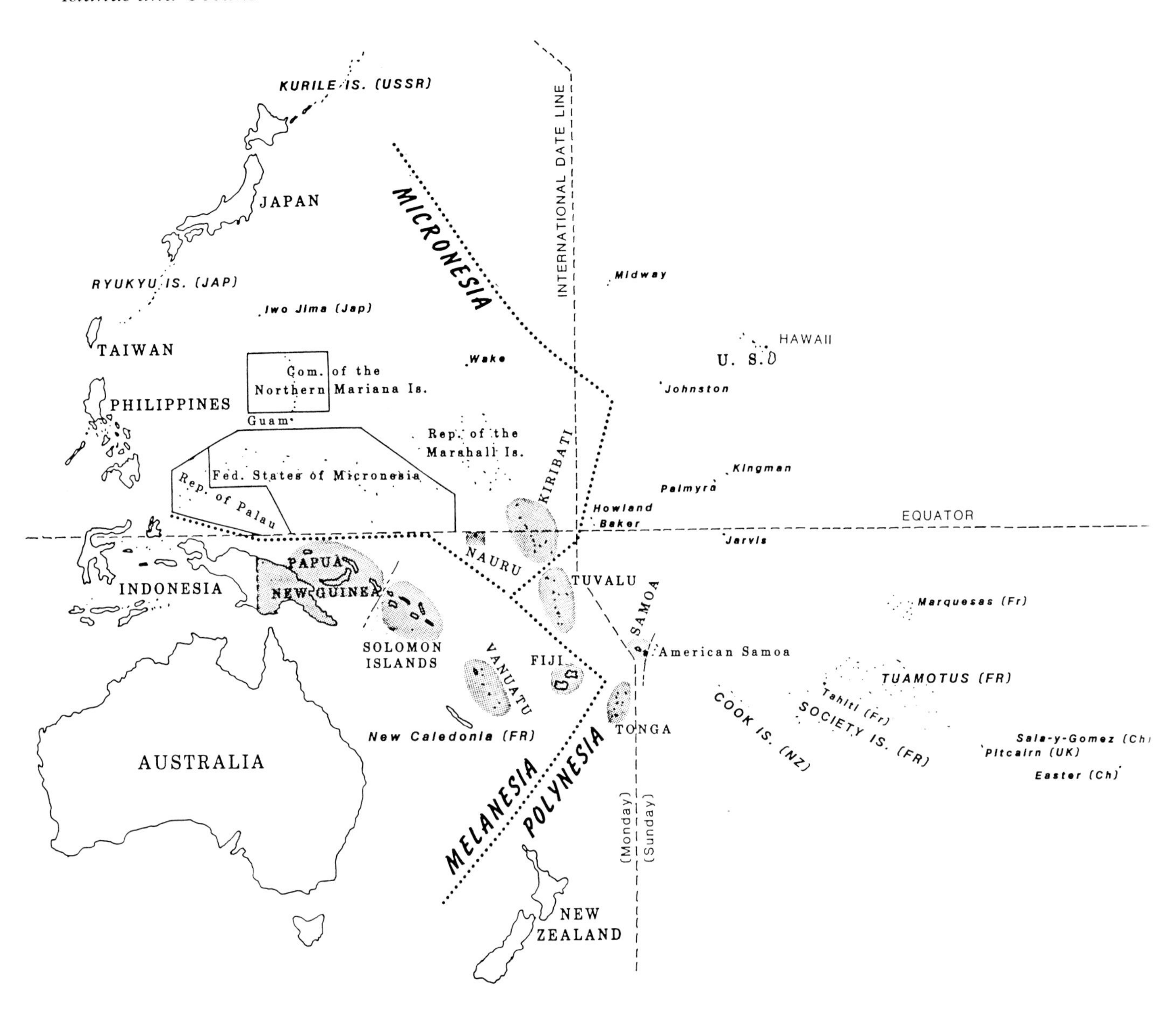

Also previously mentioned were the important Spanish and Portuguese islands (Portuguese *Madeira and Azores,* and Spanish *Canary,* p. 46).

Of the remaining Atlantic islands, all are territorial possessions of established countries, as noted below.

Island	*Affiliation*	*Location*
Bermuda	United Kingdom	E of North Carolina
Fernando de Noronha	Brazil	NE of Brazil
St. Paul's Rocks	Brazil	NE of Brazil
Trinidade	Brazil	E of Rio de Janeiro
Bouvet	Norway	SW of Cape Town
St. Pierre and Miquelon	France (Overseas Department)	SW of Newfoundland

British South Atlantic Islands (United Kingdom); all between South America and Africa

Ascension	South Georgia
St. Helena	South Sandwich
Tristan da Cunha	Falklands (also claimed by Argentina)
Gough	

III. Indian

There are four sovereign island-states in the Indian Ocean (excluding such large countries as Madagascar and Sri Lanka):

Republic of Maldives
Comoros
Seychelles
Mauritius

Maldives has already been discussed (p. 74), as has *Comoros* (p. 83). The *Seychelles* lie east of Kenya; *Mauritius* is east of Madagascar.

Minor islands of the Indian Ocean are listed below:

Islands North of Equator	*Affiliation*	*Location*
Socotra	People's Republic of Yemen	S of PR of Yemen
Laccadive	India	SW of India
Andaman	India	in Bay of Bengal

Central Islands	*Affiliation*	*Location*
Almirante	Seychelles	E of Kenya
Aldabra	Seychelles	E of Tanzania
Chagos	United Kingdom	S of India
Cocos	Australia	S of Sumatra
Christmas	Australia	S of Sumatra
Réunion	France	E of Madagascar
Cargados	Mauritius	E of Madagascar
Rodrigues	Mauritius	E of Madagascar

Islands South of Equator	*Affiliation*	*Location*
Prince Edward	Republic of South Africa	S of R.S.A.
Crozet	France	S of Madagascar
Amsterdam and St. Paul	France	SE of Madagascar
Kerguelen	France	SSE of Madagascar
McDonald and Heard	Australia	SSE of Madagascar

IV. Arctic

The larger Arctic Ocean islands (some mentioned previously) are:

Island	*Affiliation*	*Location*
Svalbard (Spitzbergen)	Norway	N of Norway
Franz Josef Land	USSR	N of USSR mainland; listed in order from W to E
Novaya Zemlya	USSR	
Severnaya Zemlya	USSR	
New Siberian Islands	USSR	
Wrangel Island	USSR	
Ellesmere	Canada	NW of Greenland
Baffin	Canada	W of Greenland
Victoria	Canada	N of central Canada
Banks	Canada	W of Victoria Island

(*Note:* There are *many* other Canadian Arctic Islands.)

This completes the section on oceanic islands. No cartograms have been provided for the Atlantic, Indian, and Arctic areas because the major islands were mentioned in earlier chapters. After an examination of the atlas, you may wish to test yourself by covering the *affiliation* and *location* columns and testing your memory of these.

Appendix

Selected Vital Statistics—1988

Name	*Area (thousands)*		*Population (Millions)*	*Years to Double Pop.*	*Density (mi²)*	*BR*	*DR*	*Urban (%)*	*Per Capita GNP 1988 (US$)*
	km²	*mi²*				*(per 1,000)*			
Afghanistan[a]	657	254	14.2	27	56	48	22	16	300
Albania[a]	29	11	3.1	34	289	26	6	34	700
Algeria[a]	2,381	920	23.5	22	26	42	10	43	2,530
Andorra[b]	< 1	< 1	0.04	65	238	14	4	85	8,000
Angola[a]	1,246	481	8.0	28	16	47	22	25	320
Antigua and Barbuda[a]	< 1	< 1	0.08	71	468	15	5	34	2,030
Argentina[a]	2,795	1,080	31.5	44	29	24	8	84	2,130
Australia[a]	7,717	2,968	16.0	86	5	16	8	86	10,840
Capital Territory	2	< 1	0.2	—	—	—	—	—	—
New South Wales	801	309	5.4	—	—	—	—	—	—
Northern Terr.	1,347	520	0.1	—	—	—	—	—	—
Queensland	1,727	667	2.5	—	—	—	—	—	—
So. Australia	984	380	1.4	—	—	—	—	—	—
Tasmania	68	26	0.4	—	—	—	—	—	—
Victoria	228	88	4.0	—	—	—	—	—	—
W. Australia	2,527	976	1.4	—	—	—	—	—	—
Austria[a]	84	32	7.6	ZPG	237	12	12	55	9,150
Azores Is. (Port.)	2	< 1	0.3	—	331	—	—	—	—
Bahamas[a]	11	4	0.2	39	50	23	6	75	7,150
Bahrain[a]	< 1	< 1	0.4	25	1,731	32	5	81	9,560
Bangladesh[a]	143	55	107.0	26	1,945	44	17	13	150
Barbados[a]	< 1	< 1	0.3	78	1,807	17	8	32	4,680
Belgium[a]	31	12	10.0	1,732	833	12	11	95	8,450
Belize[a]	23	9	0.2	26	19	33	6	52	1,130
Benin[a]	113	43	4.3	23	100	51	20	39	270
Bermuda (UK)	< 1	< 1	0.06	—	2,619	—	—	—	—
Bhutan[a]	47	18	1.5	31	83	38	18	5	160
Bolivia[a]	1,098	424	6.5	27	15	40	14	48	470
Botswana[a]	712	275	1.2	21	4	48	14	22	840
Bougainville (PNG)	10	4	0.1	—	13	—	—	—	—
Brazil[a]	8,512	3,286	142.0	33	43	29	8	71	1,640
Brunei[a]	6	2	0.2	26	90	30	4	64	17,580

Sources: U.S. Bureau of the Census; United Nations; Population Reference Bureau, Inc.
Notes: All values are rounded and do not necessarily total.
Density per km² may be obtained by density per mi² × 0.386.
ZPG = Zero Population Growth
< Less than
[a]Independent state; member of United Nations
[b]Independent state; not a member of United Nations
[c]Member of United Nations; not an independent state

Name	Area (thousands) km²	Area (thousands) mi²	Population (Millions)	Years to Double Pop.	Density (mi²)	BR (per 1,000)	DR (per 1,000)	Urban (%)	Per Capita GNP 1988 (US$)
Bulgaria[a]	111	43	9.0	578	209	13	12	66	3,520
Burkina Faso[a]	275	106	7.3	25	69	48	20	8	140
Burma[a]	678	262	33.8	33	148	34	13	24	190
Burundi[a]	29	11	5.0	24	455	47	18	5	240
Cambodia (see Kampuchea)									
Cameroun (Cameroon)[a]	475	184	10.3	26	56	43	16	42	810
Canada[a]	9,972	3,852	26.0	91	7	15	7	76	13,670
Alberta	661	255	2.3	—	—	—	—	—	—
British Columbia	948	366	2.9	—	—	—	—	—	—
Manitoba	650	251	1.1	—	—	—	—	—	—
New Brunswick	73	28	0.7	—	—	—	—	—	—
Newfoundland	404	156	0.6	—	—	—	—	—	—
Northwest Terr.	3,378	1,305	0.05	—	—	—	—	—	—
Nova Scotia	55	21	0.9	—	—	—	—	—	—
Ontario	1,068	413	9.0	—	—	—	—	—	—
Prince Edward Island	6	2	0.1	—	—	—	—	—	—
Québec	1,540	595	6.5	—	—	—	—	—	—
Saskatchewan	652	252	1.0	—	—	—	—	—	—
Yukon Territory	536	207	0.02	—	—	—	—	—	—
Canary Islands (Sp.)	7	3	1.5	—	534	—	—	—	—
Cape Verde[a]	4	2	0.3	26	209	35	8	27	430
Central African Rep.[a]	612	236	2.7	28	11	44	19	42	270
Chad[a]	1,284	496	4.6	35	9	43	23	27	150
Chile[a]	741	286	12.4	44	43	22	6	83	1,440
China (People's Rep.)[a]	9,700	3,746	1,062.0	53	284	21	8	32	310
China (Republic of)[b]	36	14	19.6	59	1,400	17	5	67	2,800
Colombia[a]	1,179	455	29.9	33	66	28	7	65	1,320
Comoros[a]	2	< 1	0.4	21	576	47	14	23	280
Congo[a]	342	132	2.1	21	16	47	13	48	1,020
Costa Rica[a]	51	20	2.8	25	140	31	4	48	1,290
Cuba[a]	114	44	10.3	58	234	18	6	71	910
Cyprus[a]	9	4	0.7	63	175	20	9	53	3,790
Czechoslovakia[a]	128	49	15.6	257	318	15	12	71	5,190
Denmark[a]	43	17	5.1	ZPG	300	11	11	84	11,240
Djibouti[a]	23	9	0.3	28	33	43	18	74	500
Dominica[a]	< 1	< 1	0.1	41	345	22	5	27	1,160
Dominican Republic[a]	49	19	6.5	28	342	33	8	52	810
Ecuador[a]	301	116	10.0	25	86	35	8	51	1,160
Egypt[a]	1,000	386	51.9	26	134	37	11	46	680
El Salvador[a]	21	8	5.3	27	662	36	10	43	710
Equatorial Guinea[a]	28	11	0.3	38	27	38	20	60	300
Ethiopia[a]	1,184	457	46.0	30	101	46	23	10	170

Faeroe Islands (Den.)	1	< 1	0.04	—	74	—	—	—	—
Falkland Islands (UK)	12	5	< 0.01	—	< 1	—	—	—	—
Federated States of Micronesia (US)	< 1	< 1	0.09	—	323	—	—	—	—
Fiji[a]	18	7	0.7	31	100	28	5	37	1,700
Finland[a]	337	130	4.9	224	38	13	10	60	10,870
Formosa (see China, Republic of)									
France[a]	551	213	55.6	178	261	14	10	73	9,550
French Guiana (Fr.)	91	35	0.08	—	2	—	—	—	—
Gabon[a]	264	102	1.2	43	12	34	18	41	3,340
Galápagos Is. (Ecuad.)	8	3	< 0.01	—	1	—	—	—	—
Gambia[a]	10	4	0.8	33	200	49	28	31	390
German Democratic Rep. (East Germany)[a]	108	42	16.7	ZPG	398	14	14	77	6,220
Germany, Fed. Rep. of (West Germany)[a]	248	96	61.0	ZPG	635	10	12	85	10,940
Ghana[a]	238	92	13.9	25	151	42	14	31	390
Gibraltar (UK)	< 1	< 1	0.03	—	12,000	—	—	100	—
Greece[a]	132	51	10.0	289	196	12	9	70	3,550
Greenland (Den.)	2,175	840	0.06	—	< 1	—	—	—	—
Grenada[a]	< 1	< 1	0.1	37	842	26	7	15	970
Guam (US)	< 1	< 1	0.1	—	572	—	—	40	—
Guatemala[a]	109	42	8.4	22	200	41	9	39	730
Guinea[a]	246	95	6.4	29	67	47	23	22	320
Guinea-Bissau[a]	36	14	0.9	35	64	41	21	27	170
Guyana[a]	215	83	0.9	35	11	26	6	32	570
Haiti[a]	28	11	6.2	30	563	36	13	26	350
Honduras[a]	112	43	4.7	22	109	39	8	40	730
Hong Kong (UK)	1	< 1	5.6	77	13,692	14	5	92	6,220
Hungary[a]	93	36	10.6	ZPG	294	12	14	56	1,940
Iceland[a]	103	40	0.2	76	5	16	7	89	10,720
India[a]	3,266	1,262	800.3	33	634	33	12	25	250
Indonesia[a]	1,905	736	174.9	33	238	31	10	22	530
Bali and Nusa Tenggara	91	35	9.6	—	274	—	—	—	—
Java and Madura	132	51	97.5	—	1,911	—	—	—	—
Kalimantan (Borneo)	539	208	7.2	—	35	—	—	—	—
Moluccas and West Irian	472	182	3.6	—	20	—	—	—	—
Sulawesi (Celebes)	189	73	12.6	—	173	—	—	—	—
Sumatra	473	183	25.2	—	138	—	—	—	—
Iran[a]	1,647	636	50.4	21	79	45	13	51	1,000
Iraq[a]	449	173	17.0	21	98	46	13	68	1,860
Ireland, Rep. of[a]	70	27	3.5	85	130	18	9	56	4,840
Israel[a]	21	8	4.4	41	550	23	7	90	4,920
Italy[a]	301	116	57.3	ZPG	494	10	10	72	6,520
Ivory Coast[a]	330	128	10.8	23	84	46	15	43	620
Jamaica[a]	11	4	2.5	34	625	26	5	54	940
Japan[a]	370	143	122.2	124	855	12	6	76	11,330
Jordan[a]	97	38	3.7	19	97	45	8	60	1,560
Kampuchea[a]	181	70	6.5	33	93	39	18	11	240
Kenya[a]	582	225	22.4	18	100	52	13	16	290
Kiribati[b]	< 1	< 1	0.06	23	227	37	9	20	500

Name	Area (thousands) km²	Area (thousands) mi²	Population (Millions)	Years to Double Pop.	Density (mi²)	BR (per 1,000)	DR (per 1,000)	Urban (%)	Per Capita GNP 1988 (US$)
Korea, Dem. Rep. of (North Korea)[b]	120	47	21.4	28	455	30	5	64	840
Korea, Rep. of (South Korea)[b]	98	38	42.1	51	1,108	20	6	65	2,180
Kuwait[a]	16	6	1.9	22	317	34	3	80	14,270
Laos[a]	237	91	3.8	28	42	41	16	16	150
Lebanon[a]	10	4	3.3	32	825	30	8	80	1,200
Lesotho[a]	30	12	1.6	27	133	41	15	17	480
Liberia[a]	111	43	2.4	22	56	48	16	40	470
Libya[a]	1,759	679	3.8	23	6	39	9	76	7,500
Liechtenstein[b]	< 1	< 1	0.02	300	432	15	7	85	16,500
Luxembourg[a]	3	1	0.4	ZPG	400	11	11	78	13,380
Macau (Port.)	< 1	< 1	0.4	41	50,000	23	6	100	—
Madagascar[a]	590	228	10.6	25	46	44	16	22	250
Madeira Islands (Port.)	< 1	< 1	0.3	—	977	—	—	—	—
Malawi[a]	93	36	7.4	22	206	53	21	12	170
Malaysia[a]	333	128	16.1	28	126	31	7	32	2,050
Maldives[a]	< 1	< 1	0.2	18	1,577	48	10	26	290
Mali[a]	1,201	464	8.4	24	18	51	22	18	140
Malta[a]	< 1	< 1	0.4	89	2,730	16	8	85	3,300
Marshall Islands, Rep. of (US)	< 1	< 1	0.04	—	500	—	—	—	—
Mauritania[a]	1,030	398	2.0	23	5	50	20	35	410
Mauritius[a]	2	< 1	1.1	57	1,280	19	7	42	1,070
Mexico[a]	1,963	758	81.9	28	108	31	7	70	2,080
Monaco[b]	< 1	< 1	0.03	ZPG	50,000	20	20	100	17,000
Mongolia[a]	1,564	604	2.0	26	3	37	11	51	900
Morocco[a]	634	245	24.4	27	100	36	10	43	610
Mozambique[a]	771	298	14.7	27	49	45	19	13	160
Namibia (So. Africa)	824	318	1.3	21	4	44	11	51	1,150
Nauru[b]	< 1	< 1	0.01	27	1,000	34	9	85	20,000
Nepal[a]	141	54	17.8	28	330	42	17	7	160
Netherlands[a]	41	16	14.6	182	912	12	9	89	16,380
New Caledonia (Fr.)	17	6	0.1	—	23	—	—	—	—
New Zealand[a]	269	104	3.3	92	32	16	8	84	7,310
Nicaragua[a]	140	54	3.5	20	65	43	9	53	850
Niger[a]	1,267	489	7.0	24	14	51	22	16	200
Nigeria[a]	923	357	108.6	25	304	46	18	28	370
Northern Mariana Islands (US)	< 1	< 1	0.02	—	99	—	—	—	—
Norway[a]	324	125	4.2	408	34	12	11	70	13,890
Oman[a]	212	82	1.3	21	16	47	14	9	7,080
Pakistan[a]	804	310	104.6	24	337	44	15	28	380
Palau, Rep. of (US)	< 1	< 1	0.01	—	73	—	—	—	—
Panama[a]	76	29	2.3	32	79	27	5	51	2,020
Papua New Guinea[a]	475	183	3.6	29	20	36	12	13	710
Paraguay[a]	407	157	4.3	24	27	36	7	43	940

Peru[a]	1,285	496	20.7	28	42	35	10	69	960
Philippines[a]	300	116	61.5	25	530	35	7	40	600
Poland[a]	312	121	37.8	88	312	18	10	60	2,120
Portugal[a]	92	36	10.3	257	286	12	10	30	1,970
Puerto Rico (US)	9	3	3.3	54	955	19	7	67	4,850
Qatar[a]	21	8	0.3	23	37	34	4	86	15,980
Romania[a]	237	92	22.9	141	249	16	11	53	1,900
Rwanda[a]	26	10	6.8	19	680	53	16	6	290
St. Christopher (St. Kitts) and Nevis[a]	< 1	< 1	0.05	45	465	26	11	45	1,520
St. Lucia[a]	< 1	< 1	0.1	28	513	30	6	40	1,210
St. Vincent and the Grenadines[a]	< 1	< 1	0.1	35	680	26	7	25	840
Samoa[a]	3	1	0.2	29	146	31	7	21	660
Samoa (US)	< 1	< 1	0.03	—	441	—	—	—	—
San Marino[b]	< 1	< 1	0.02	194	943	10	7	93	4,400
São Tomé and Príncipe[a]	1	< 1	0.1	25	282	36	9	35	310
Saudi Arabia[a]	2,252	870	14.8	22	17	39	7	72	8,860
Senegal[a]	196	76	7.1	24	93	46	18	36	370
Seychelles[a]	< 1	< 1	0.1	36	386	27	7	37	1,160
Sierra Leone[a]	72	28	3.9	38	139	47	29	28	370
Singapore[a]	< 1	< 1	2.6	61	11,607	17	5	100	7,420
Solomon Islands[a]	30	12	0.3	19	25	42	6	9	510
Somalia[a]	637	246	7.7	28	31	48	23	34	270
South Africa[a]	1,222	472	34.3	30	73	33	10	56	2,010
Spain[a]	505	195	39.0	147	200	13	8	70	4,360
Sri Lanka[a]	66	25	16.3	38	652	25	7	22	370
Sudan[a]	2,505	968	23.5	24	24	45	16	20	330
Suriname[a]	143	55	0.4	33	7	27	7	66	2,570
Swaziland[a]	17	7	0.7	22	100	47	16	26	650
Sweden[a]	450	174	8.4	636	48	12	11	83	11,890
Switzerland[b]	41	16	6.6	301	412	12	9	57	16,380
Syria[a]	187	72	11.3	18	157	47	9	49	1,630
Taiwan (see China, Republic of)									
Tanzania[a]	942	364	23.5	20	65	50	15	18	270
Thailand[a]	518	200	53.6	33	268	29	8	17	830
Togo[a]	57	22	3.2	22	145	47	15	22	250
Tonga[b]	1	< 1	0.1	28	381	27	3	40	776
Trinidad and Tobago[a]	5	2	1.3	34	650	27	7	34	6,010
Tunisia[a]	164	63	7.6	27	121	32	7	53	1,220
Turkey[a]	768	297	51.4	33	173	30	9	46	1,130
Tuvalu[b]	< 1	< 1	0.01	27	927	35	8	35	700
Uganda[a]	243	94	15.9	20	169	50	16	10	200
Union of Soviet Socialist Republics[a]	22,272	8,599	285.0	79	33	19	11	65	7,400
Armenian SSR	30	12	3.3	—	—	—	—	—	—
Azerbaydzhan SSR	87	33	6.5	—	—	—	—	—	—
Belorussian SSR[c]	208	80	10.0	—	—	—	—	—	—
Estonian SSR	45	17	1.7	—	—	—	—	—	—
Georgian SSR	70	27	5.6	—	—	—	—	—	—
Kazakh SSR	2,715	1,048	15.9	—	—	—	—	—	—

Name	Area (thousands)		Population (Millions)	Years to Double Pop.	Density (mi²)	BR (per 1,000)	DR (per 1,000)	Urban (%)	Per Capita GNP 1988 (US$)
	km²	mi²							
Kirgiz SSR	199	77	3.9	—	—	—	—	—	—
Latvian SSR	64	25	2.8	—	—	—	—	—	—
Lithuanian SSR	65	25	3.6	—	—	—	—	—	—
Moldavian SSR	34	13	4.4	—	—	—	—	—	—
Russian SFSR	17,075	6,591	148.0	—	—	—	—	—	—
Tadzhik SSR	143	55	4.4	—	—	—	—	—	—
Turkmen SSR	488	188	3.1	—	—	—	—	—	—
Ukrainian SSR[c]	601	232	53.0	—	—	—	—	—	—
Uzbek SSR	450	174	17.5	—	—	—	—	—	—
United Arab Emirates[a]	84	32	1.4	27	44	30	4	81	19,120
United Kingdom[a]	244	94	56.8	462	604	13	12	90	8,390
England	130	50	47.0	—	—	—	—	—	—
Channel Islands (dependency)	< 1	< 1	0.13	—	—	—	—	—	—
Isle of Man (dependency)	< 1	< 1	0.06	—	—	—	—	—	—
Northern Ireland	14	5	1.6	—	—	—	—	—	—
Scotland	79	30	5.4	—	—	—	—	—	—
Wales	21	8	2.8	—	—	—	—	—	—
United States[a]	9,520	3,676	245.0	102	67	16	9	74	16,400
Alabama	134	52	4.09	—	—	—	—	—	—
Alaska	1,518	586	0.60	—	—	—	—	—	—
Arizona	295	114	3.46	—	—	—	—	—	—
Arkansas	137	53	2.40	—	—	—	—	—	—
California	411	159	27.89	—	—	—	—	—	—
Colorado	270	104	3.42	—	—	—	—	—	—
Connecticut	13	5	3.21	—	—	—	—	—	—
Delaware	5	2	0.64	—	—	—	—	—	—
District of Columbia	< 1	< 1	0.62	—	—	—	—	—	—
Florida	152	59	12.32	—	—	—	—	—	—
Georgia	152	59	6.26	—	—	—	—	—	—
Hawaii	17	6	1.10	—	—	—	—	—	—
Idaho	216	84	1.04	—	—	—	—	—	—
Illinois	146	56	11.59	—	—	—	—	—	—
Indiana	94	36	5.50	—	—	—	—	—	—
Iowa	146	56	2.87	—	—	—	—	—	—
Kansas	213	82	2.50	—	—	—	—	—	—
Kentucky	105	40	3.76	—	—	—	—	—	—
Louisiana	126	49	4.63	—	—	—	—	—	—
Maine	86	33	1.18	—	—	—	—	—	—
Maryland	27	11	4.48	—	—	—	—	—	—
Massachusetts	21	8	5.87	—	—	—	—	—	—
Michigan	151	58	9.00	—	—	—	—	—	—
Minnesota	218	84	4.25	—	—	—	—	—	—
Mississippi	124	48	2.66	—	—	—	—	—	—

Missouri	180	70	5.11	—	—	—	—	—	—
Montana	381	147	0.85	—	—	—	—	—	—
Nebraska	200	77	1.62	—	—	—	—	—	—
Nevada	286	111	1.02	—	—	—	—	—	—
New Hampshire	24	9	1.04	—	—	—	—	—	—
New Jersey	20	8	7.66	—	—	—	—	—	—
New Mexico	315	122	1.53	—	—	—	—	—	—
New York	128	50	17.90	—	—	—	—	—	—
North Carolina	136	53	6.46	—	—	—	—	—	—
North Dakota	183	71	0.70	—	—	—	—	—	—
Ohio	107	41	10.72	—	—	—	—	—	—
Oklahoma	181	70	3.45	—	—	—	—	—	—
Oregon	251	97	2.72	—	—	—	—	—	—
Pennsylvania	117	45	11.85	—	—	—	—	—	—
Rhode Island	3	1	1.07	—	—	—	—	—	—
South Carolina	80	31	3.47	—	—	—	—	—	—
South Dakota	199	77	0.72	—	—	—	—	—	—
Tennessee	109	42	4.85	—	—	—	—	—	—
Texas	692	267	17.60	—	—	—	—	—	—
Utah	220	85	1.75	—	—	—	—	—	—
Vermont	25	10	0.55	—	—	—	—	—	—
Virginia	106	41	5.90	—	—	—	—	—	—
Washington	177	68	4.56	—	—	—	—	—	—
West Virginia	63	24	1.93	—	—	—	—	—	—
Wisconsin	145	56	4.82	—	—	—	—	—	—
Wyoming	253	99	0.54	—	—	—	—	—	—
Uruguay[a]	187	72	3.1	90	43	18	10	85	1,660
Vanuatu[a]	16	6	0.2	21	23	39	5	18	700
Vatican City[b]	< 1	< 1	< 0.01	ZPG	5,882	NA	NA	100	—
Venezuela[a]	912	352	18.3	26	52	32	6	76	3,110
Vietnam[a]	329	127	62.2	27	490	34	8	19	200
Virgin Islands (US)	< 1	< 1	0.1	—	814	—	—	—	—
Yemen, People's Democratic Republic of[a]	290	112	2.4	23	21	47	17	40	540
Yemen Arab Republic[a]	195	75	6.5	20	87	53	19	15	520
Yugoslavia[a]	256	99	23.4	102	236	16	9	46	2,070
Zaïre[a]	2,344	905	31.8	23	35	45	15	34	170
Zambia[a]	752	291	7.1	20	24	50	15	43	400
Zimbabwe[a]	389	150	9.4	20	63	47	12	24	650

Practice Maps

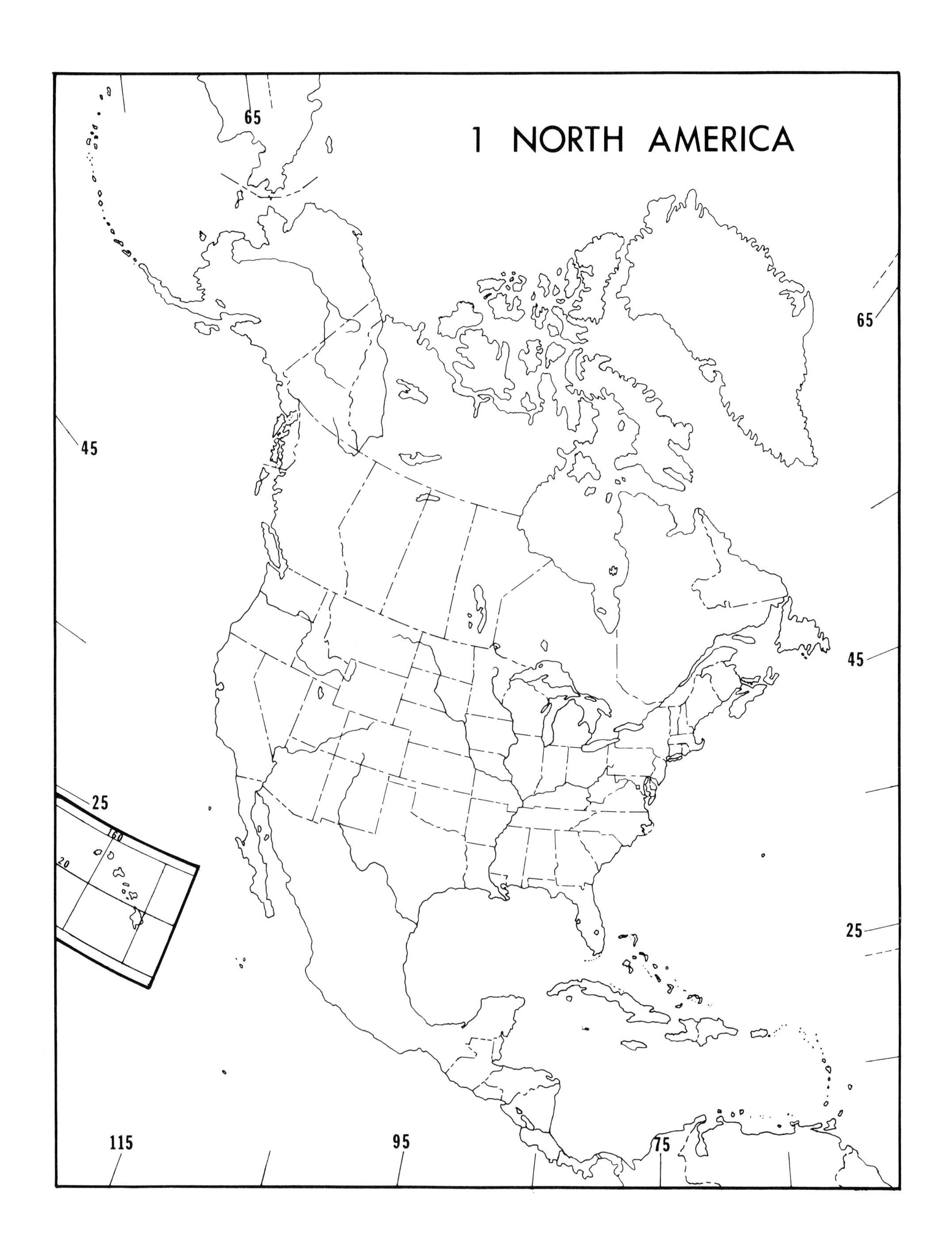
1 NORTH AMERICA
65
65
45
45
25
25
160
20
115
95
75

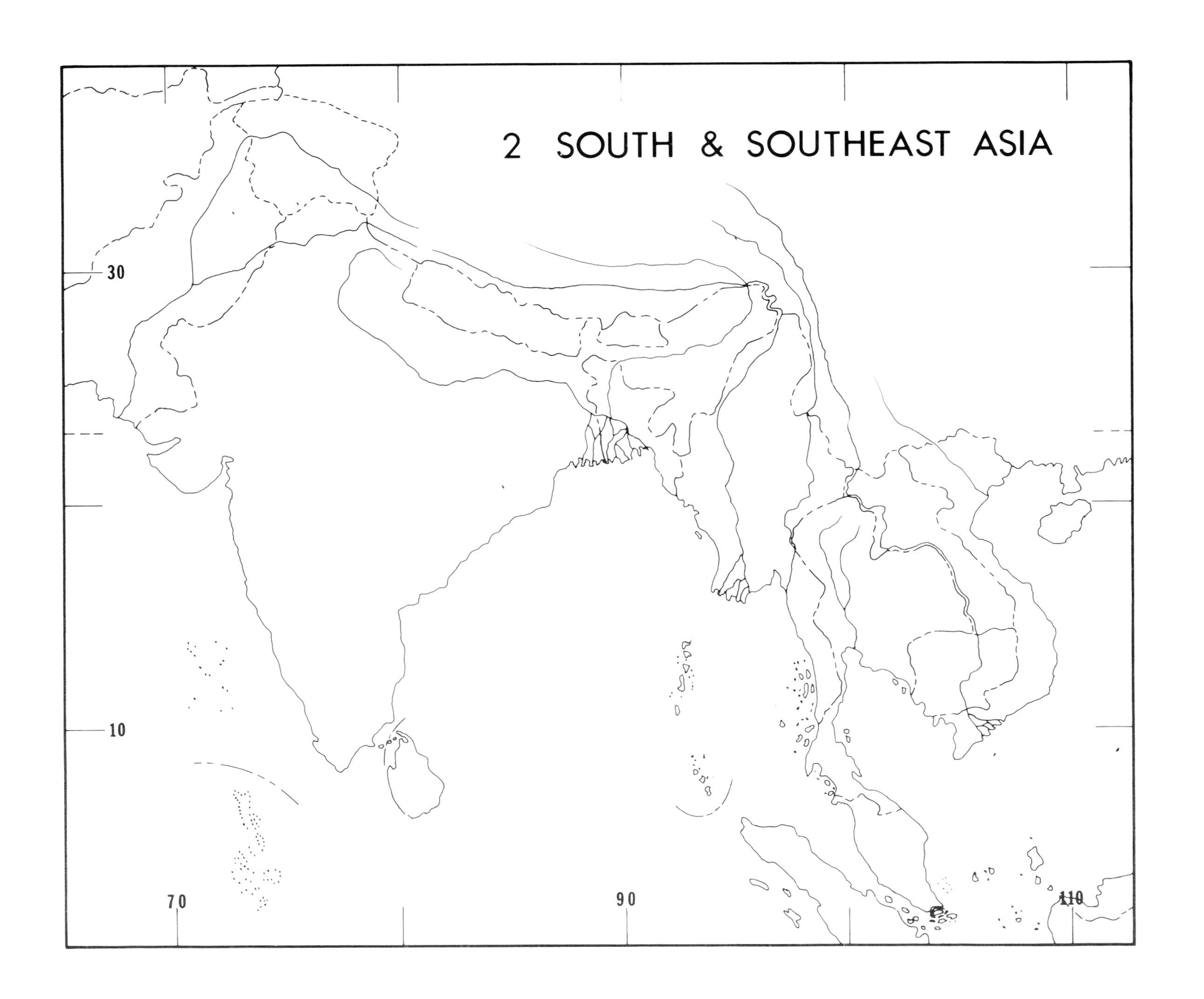
2 SOUTH & SOUTHEAST ASIA
30
10
70
90
110

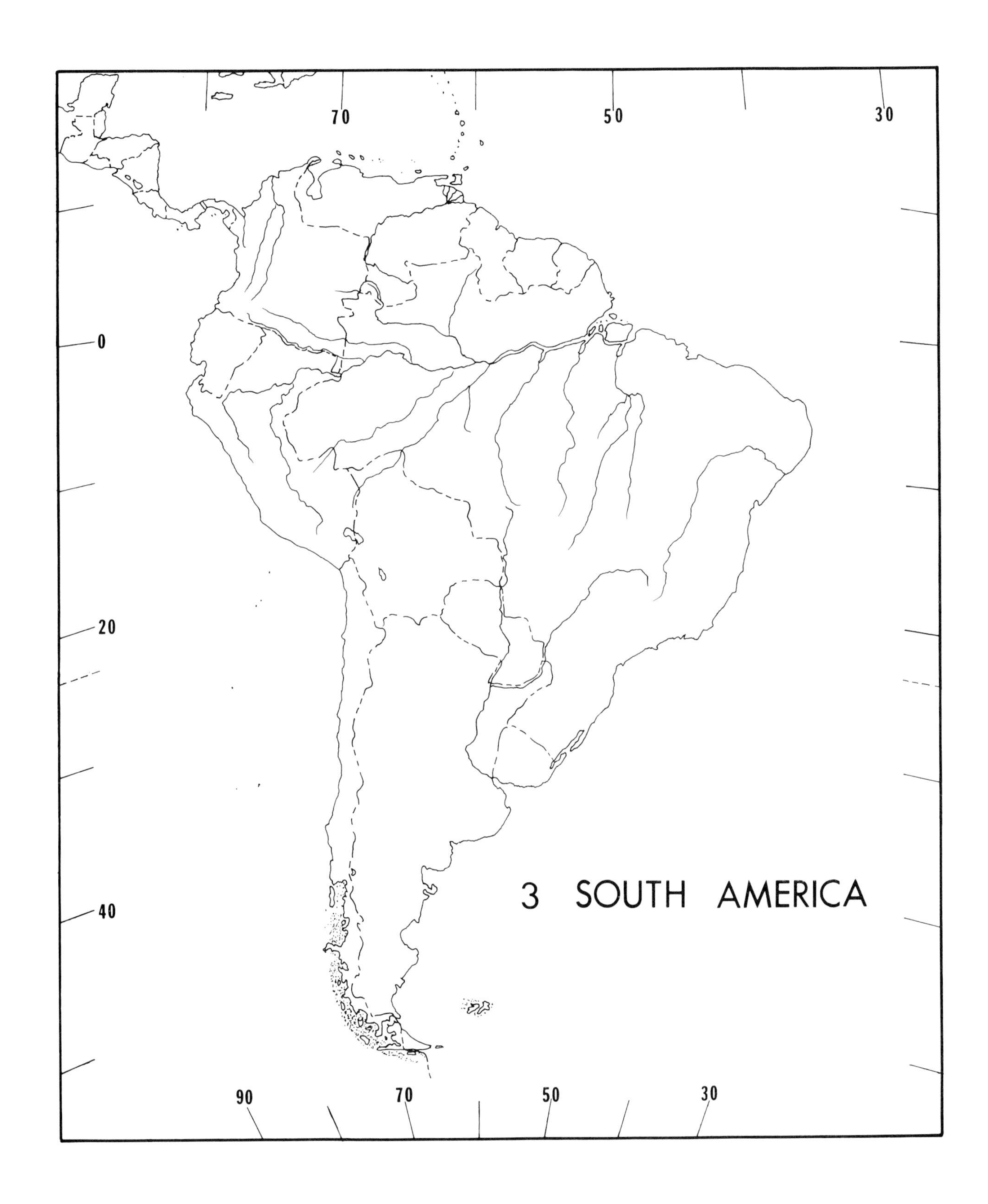
70
50
30
0
20
40
90
70
50
30
3 SOUTH AMERICA

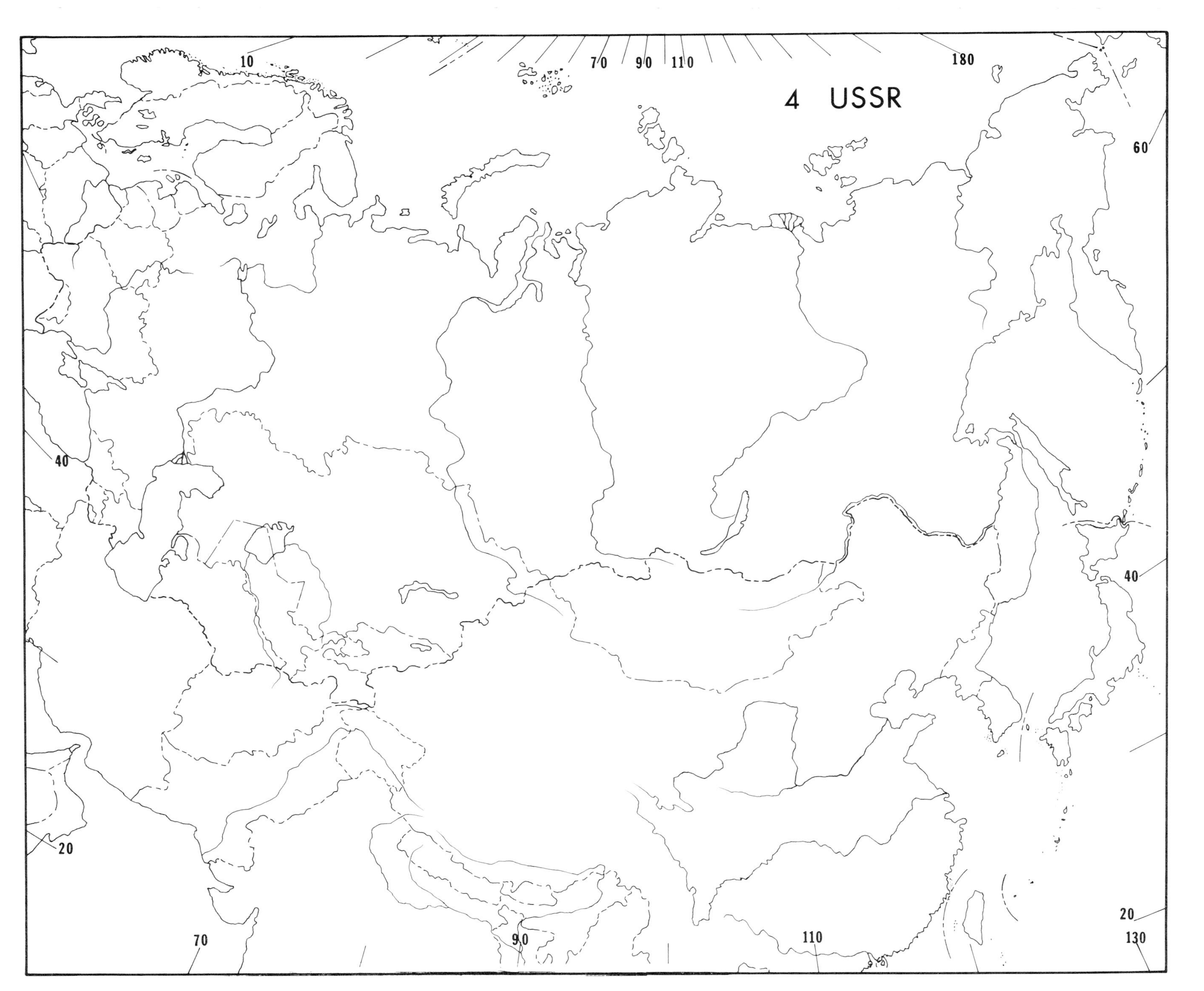
4 USSR
10
70
90
110
180
60
40
40
20
20
130
70
90
110

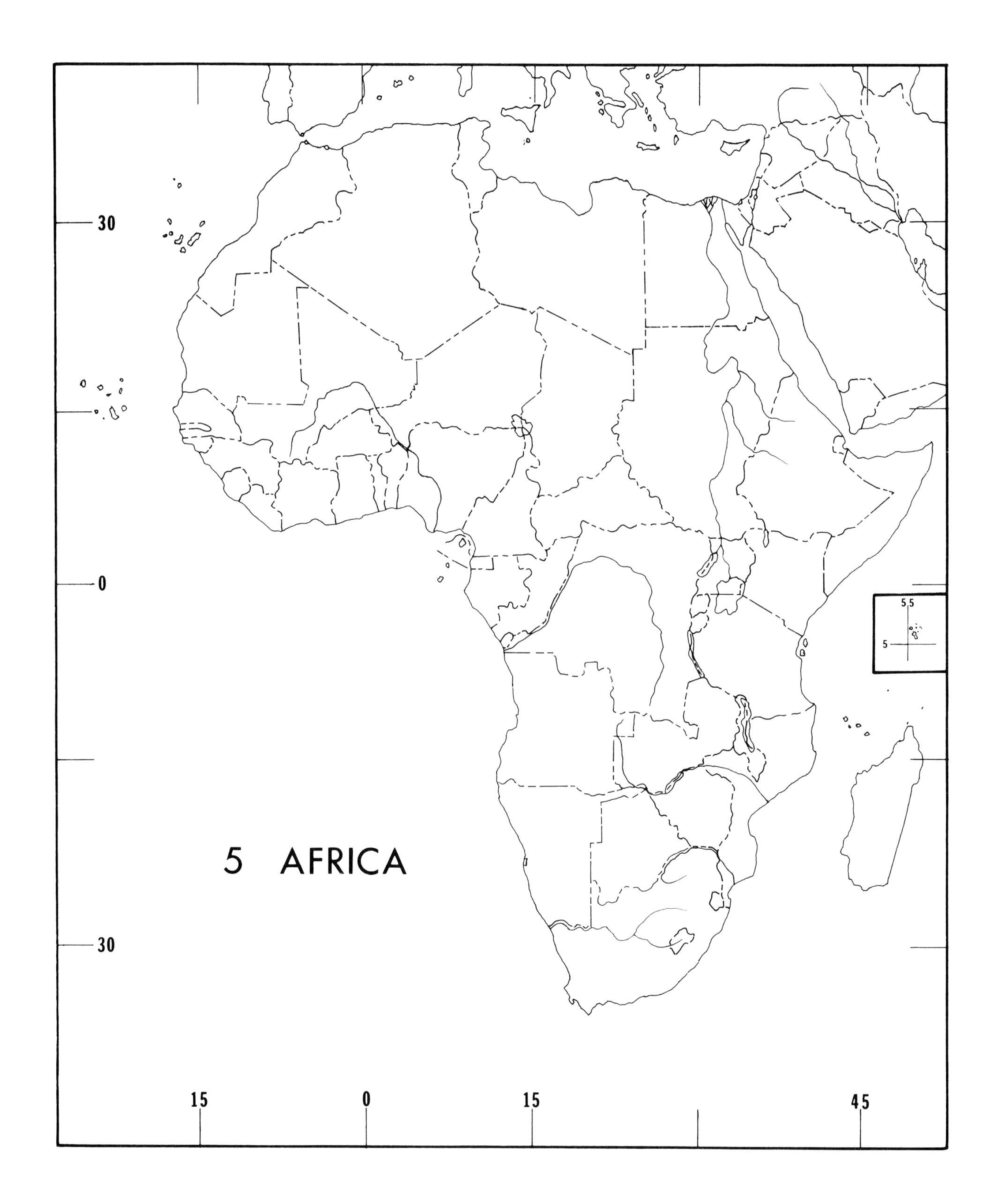
30
0
30
15
0
15
45
5 5
5
5 AFRICA

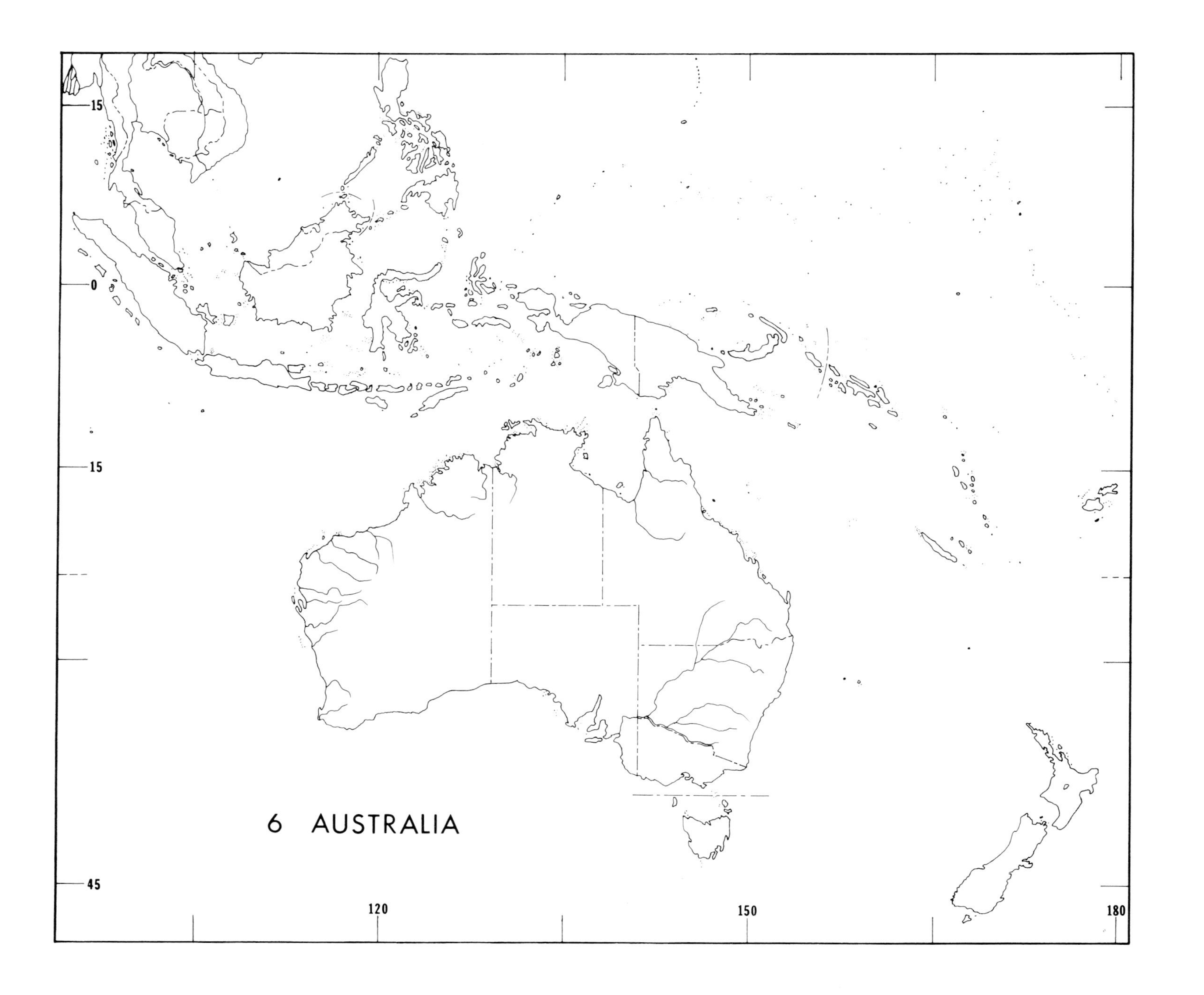
15
0
15
45
120
150
180
6 AUSTRALIA

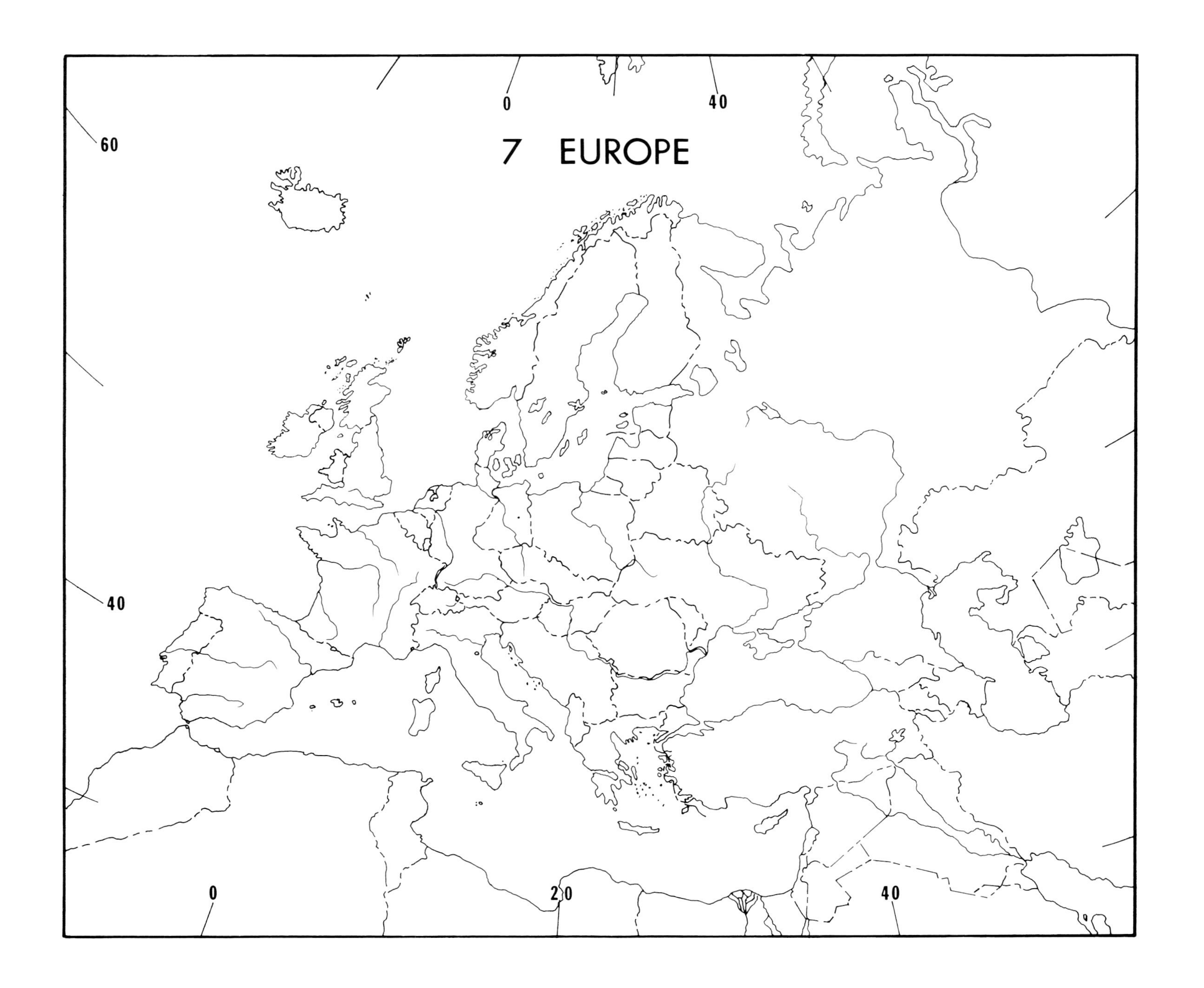
0
40
60
7 EUROPE
40
0
20
40

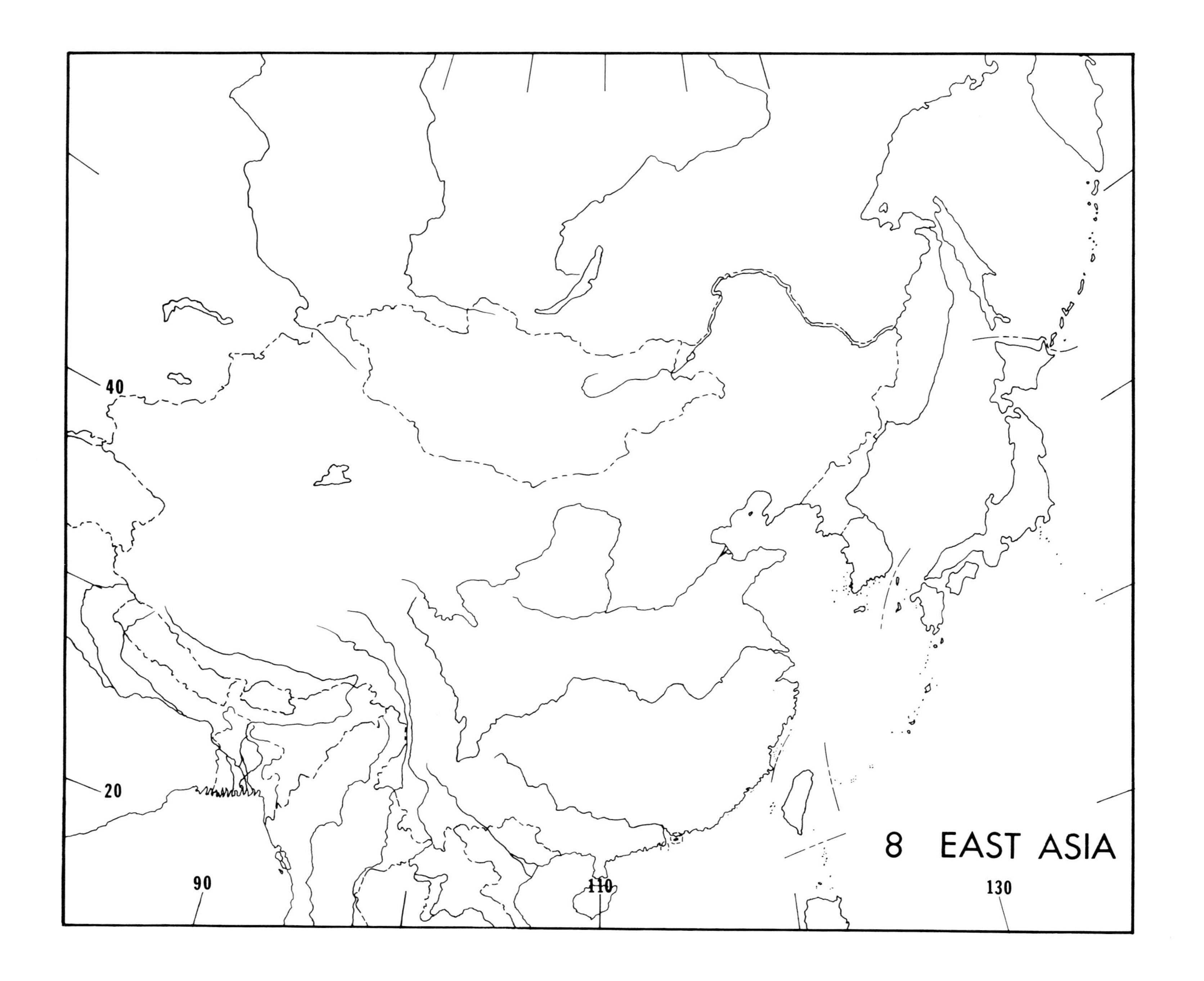
8 EAST ASIA
40
20
90
110
130

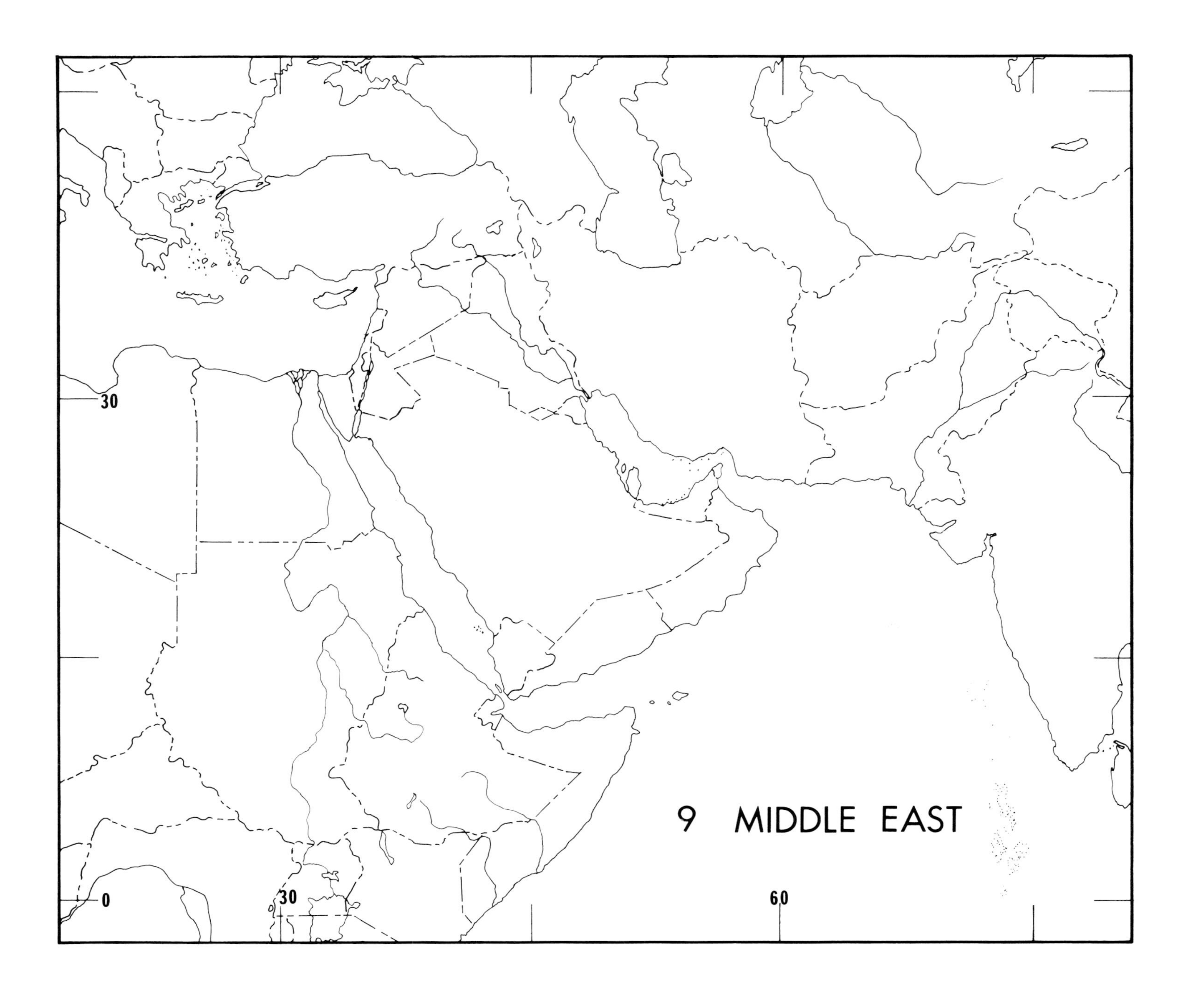
30
0
30
60
9 MIDDLE EAST

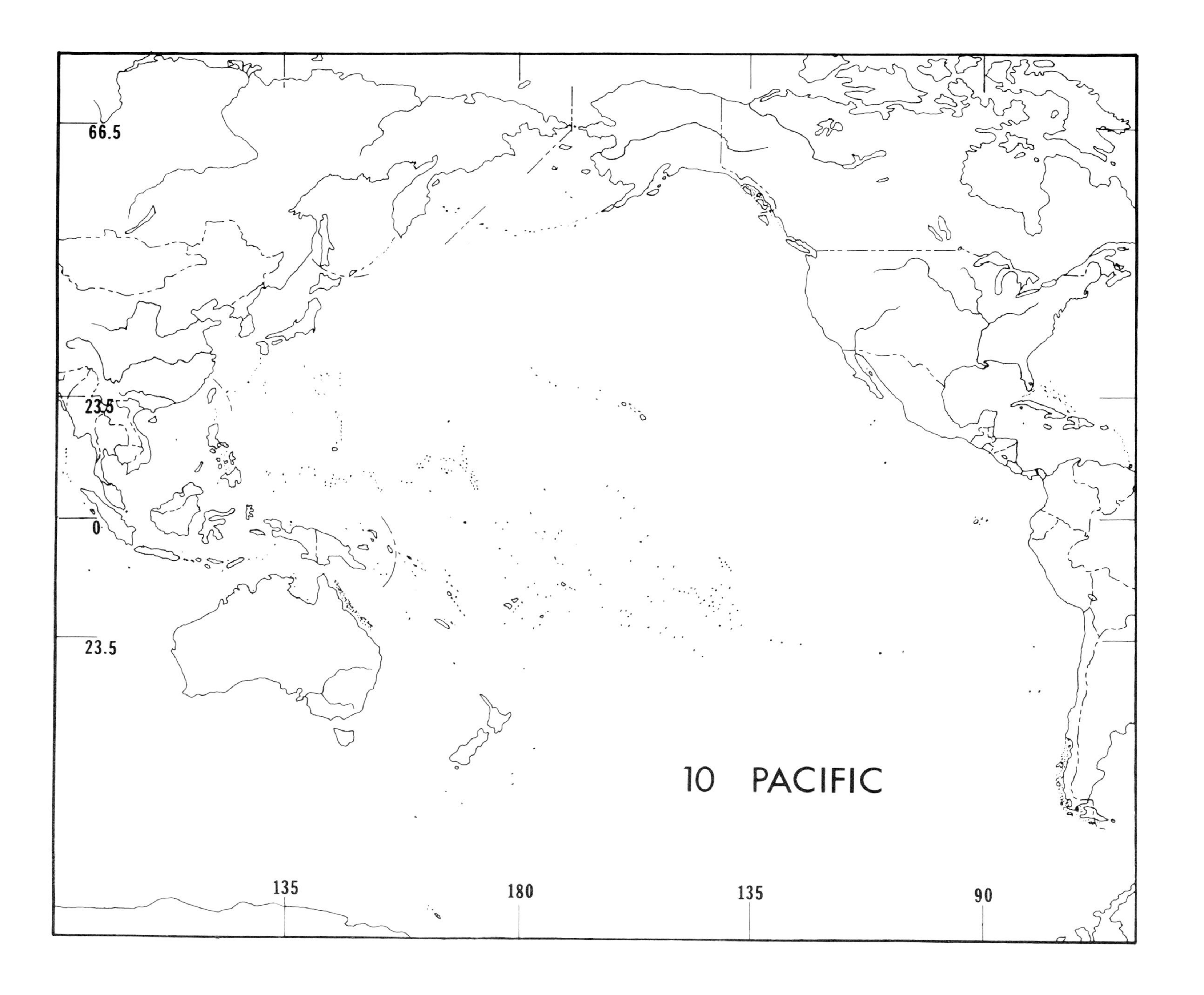
66.5
23.5
0
23.5
10 PACIFIC
135
180
135
90

Test Maps

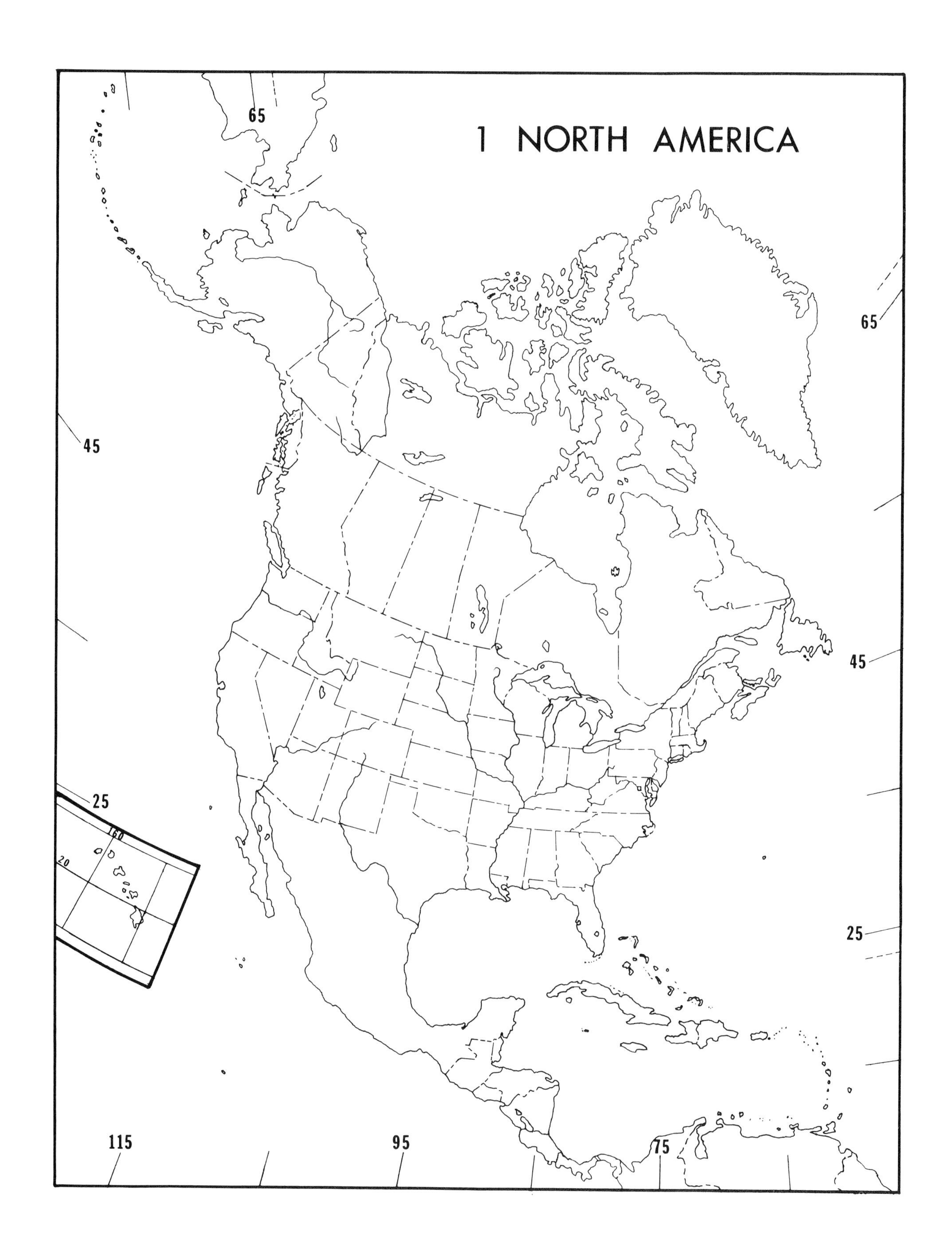
1 NORTH AMERICA
65
65
45
45
25
25
160
20
115
95
75

2 SOUTH & SOUTHEAST ASIA

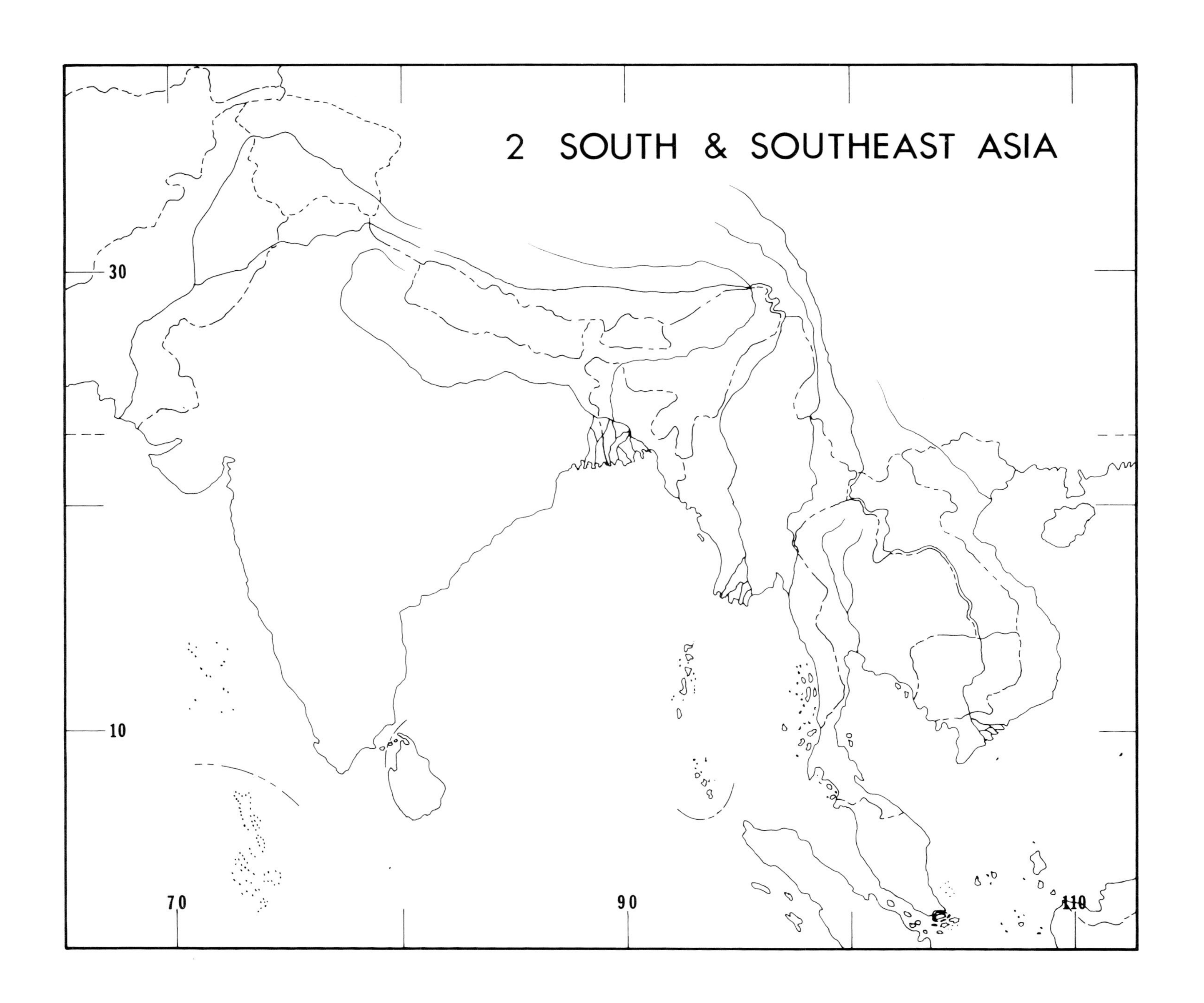

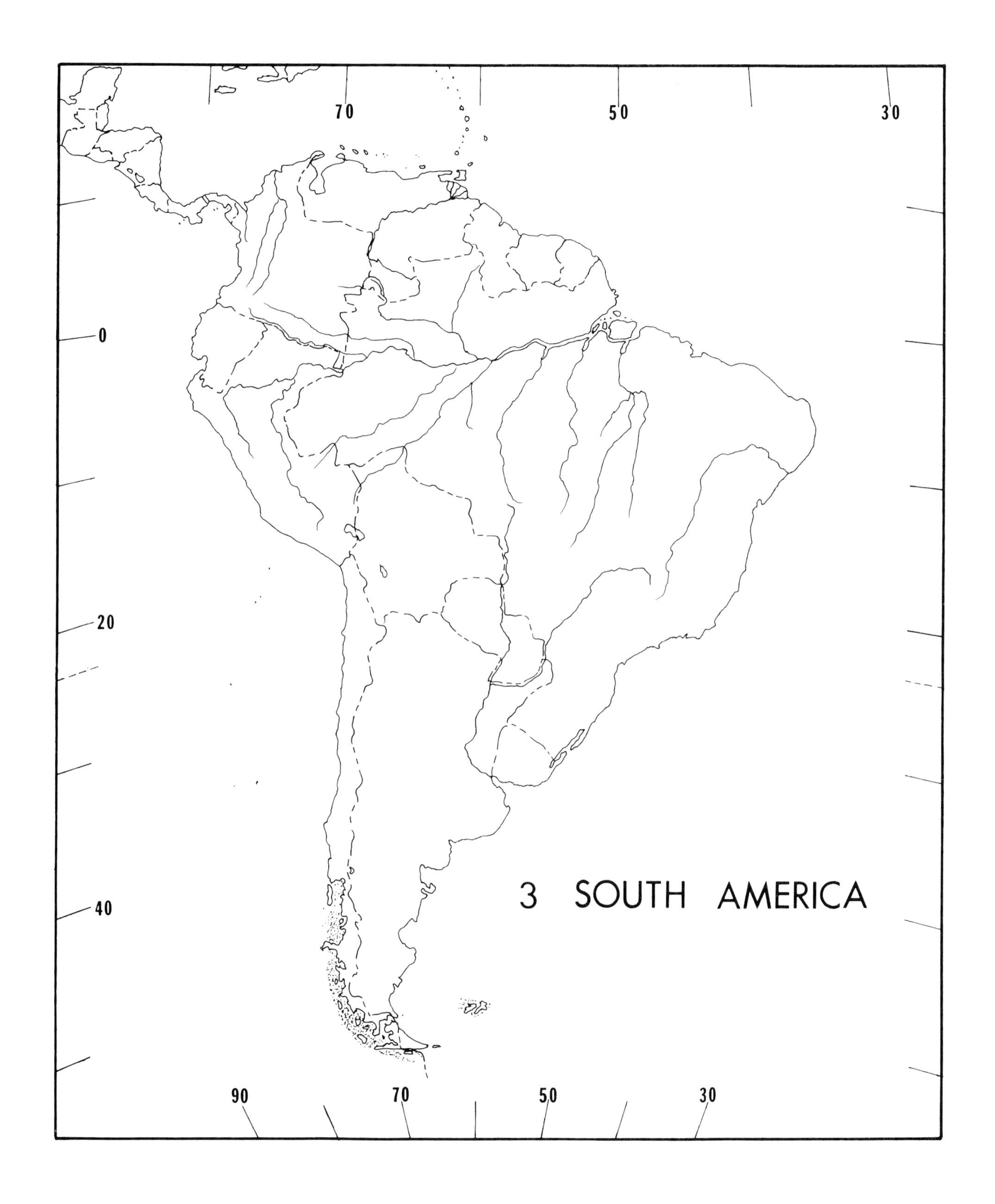

70
50
30
0
20
40
90
70
50
30
3 SOUTH AMERICA

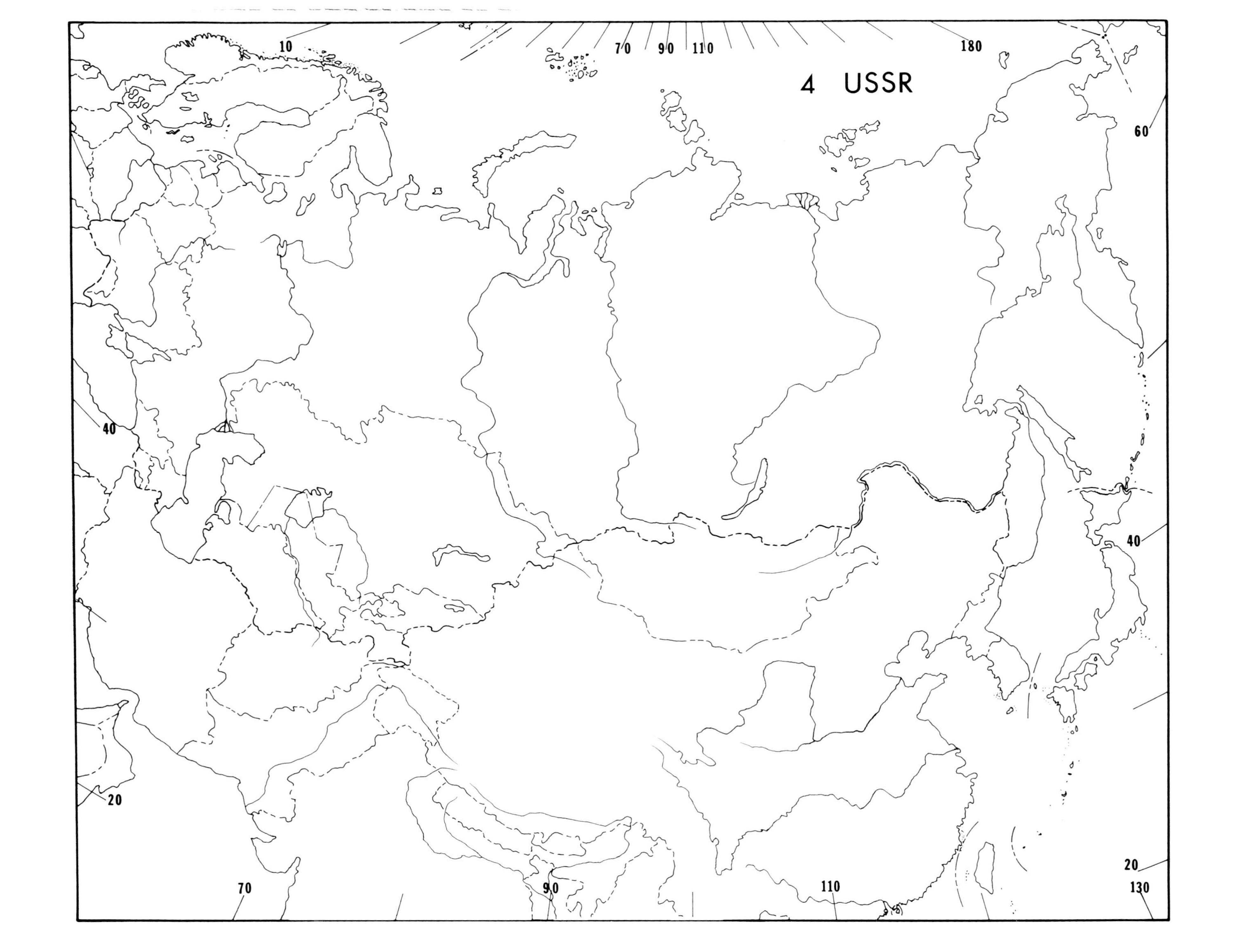
4 USSR
10
70
90
110
180
60
40
40
20
20
70
90
110
130

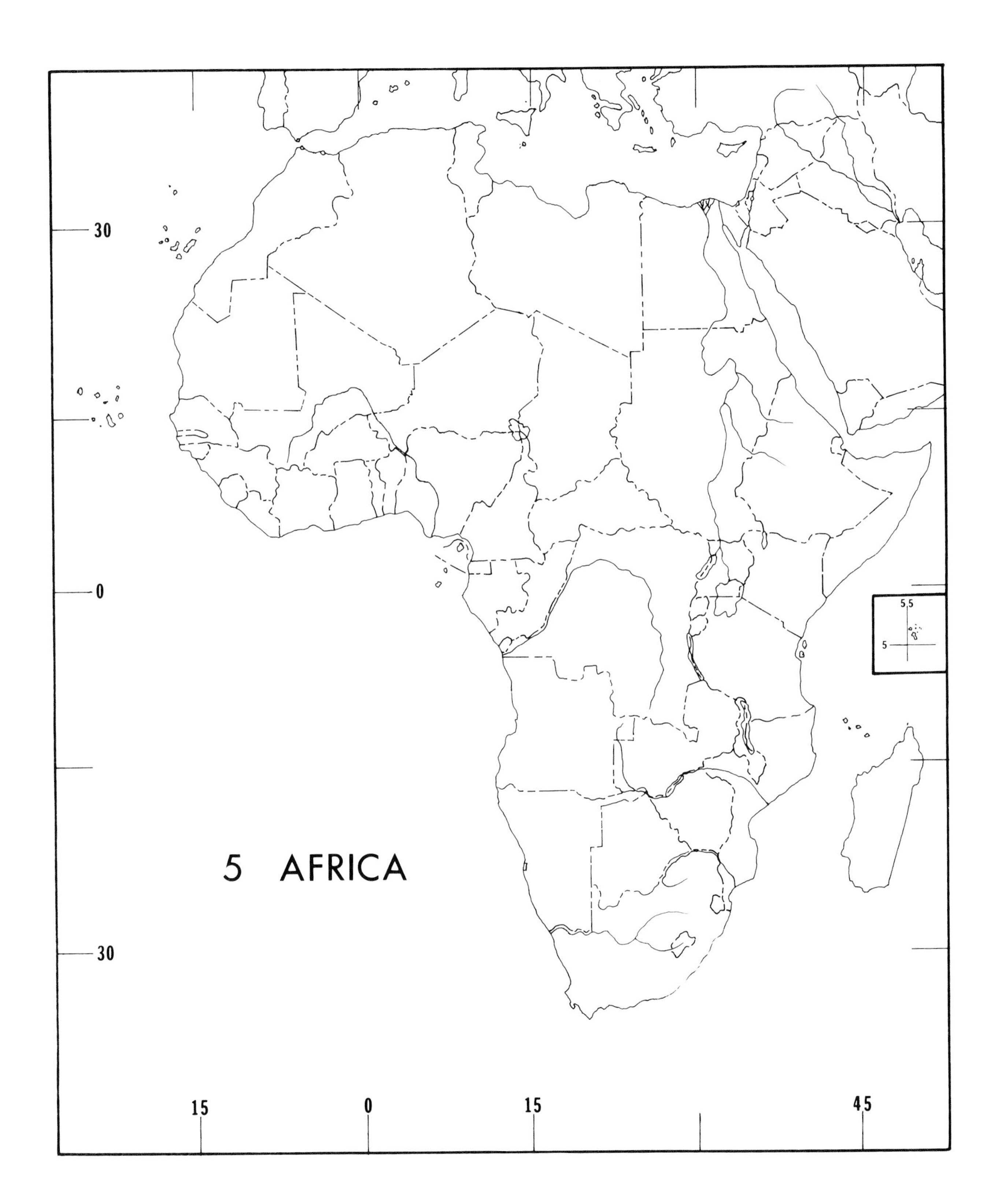
30
0
30
15
0
15
45
55
5
5 AFRICA

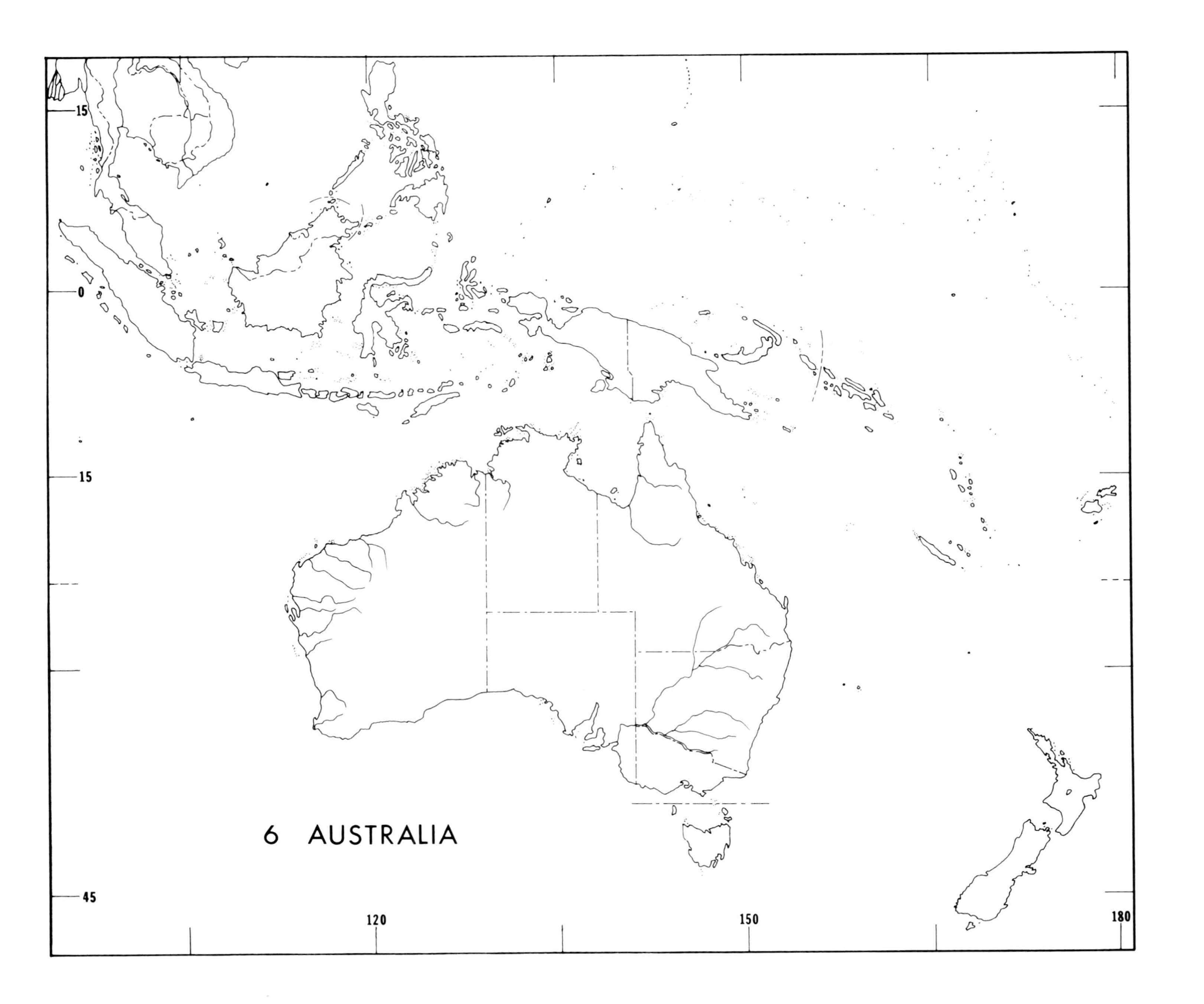
15
0
15
45
120
150
180
6 AUSTRALIA

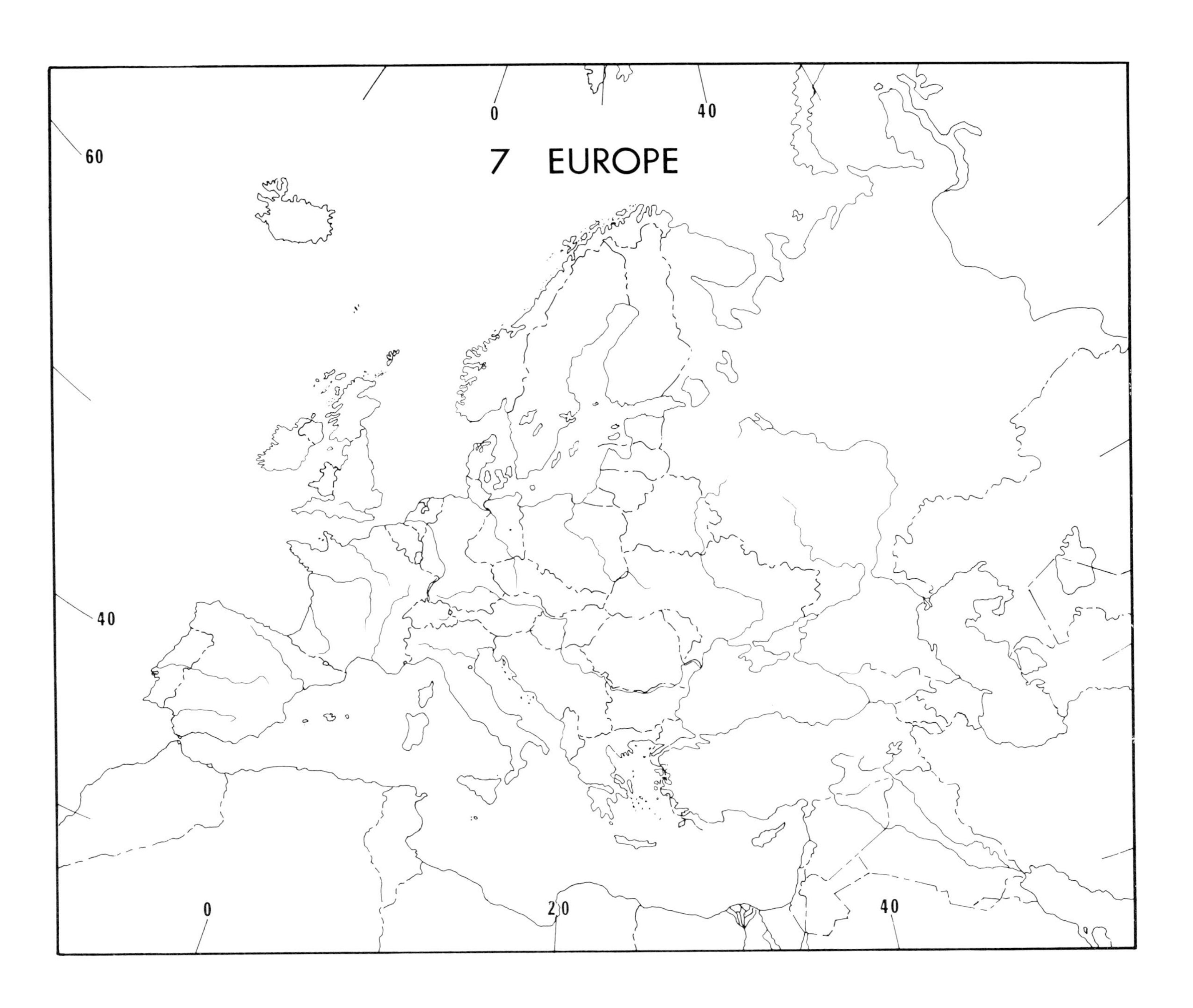

7 EUROPE
0
40
60
40
0
20
40

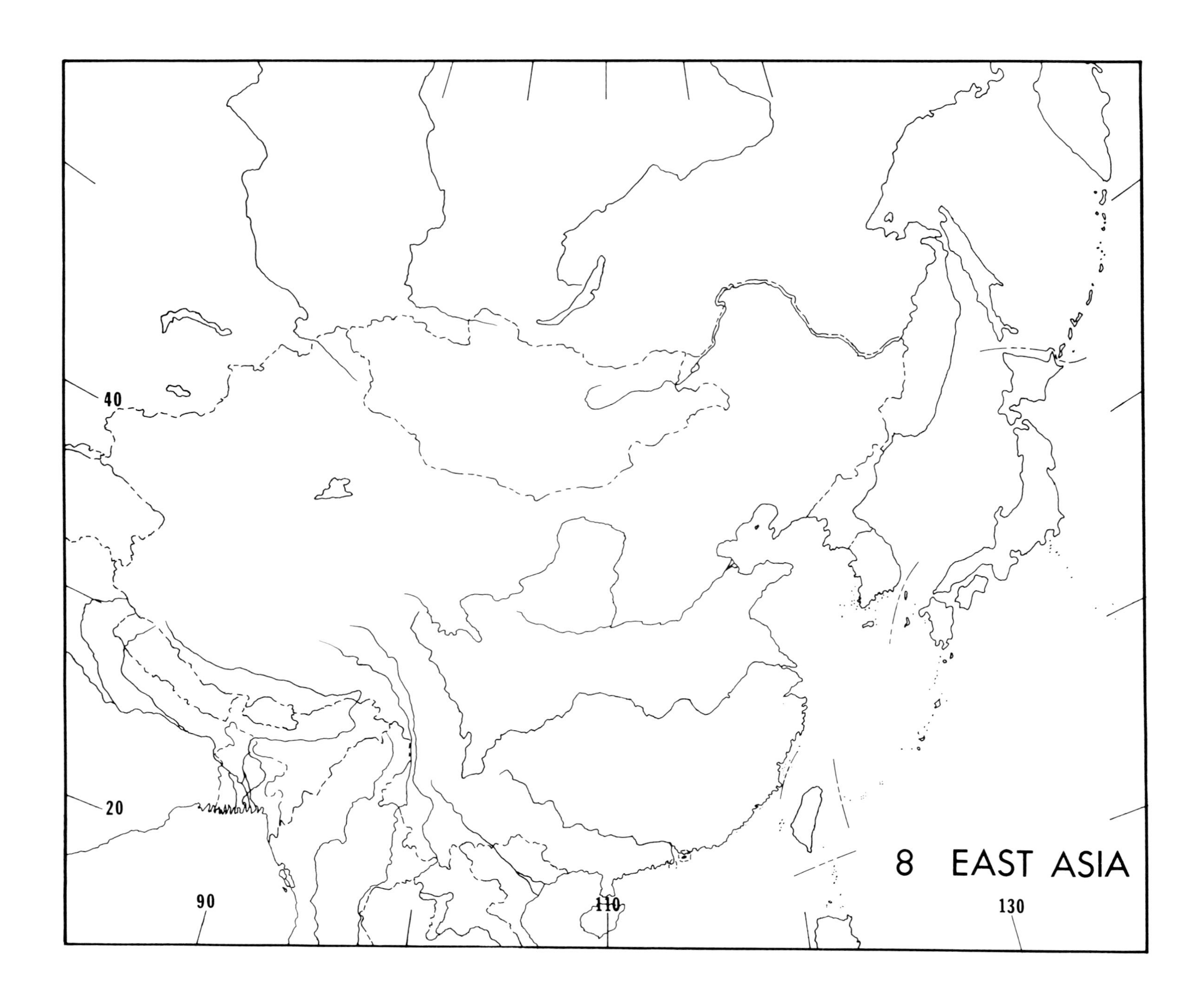
8 EAST ASIA
40
20
90
110
130

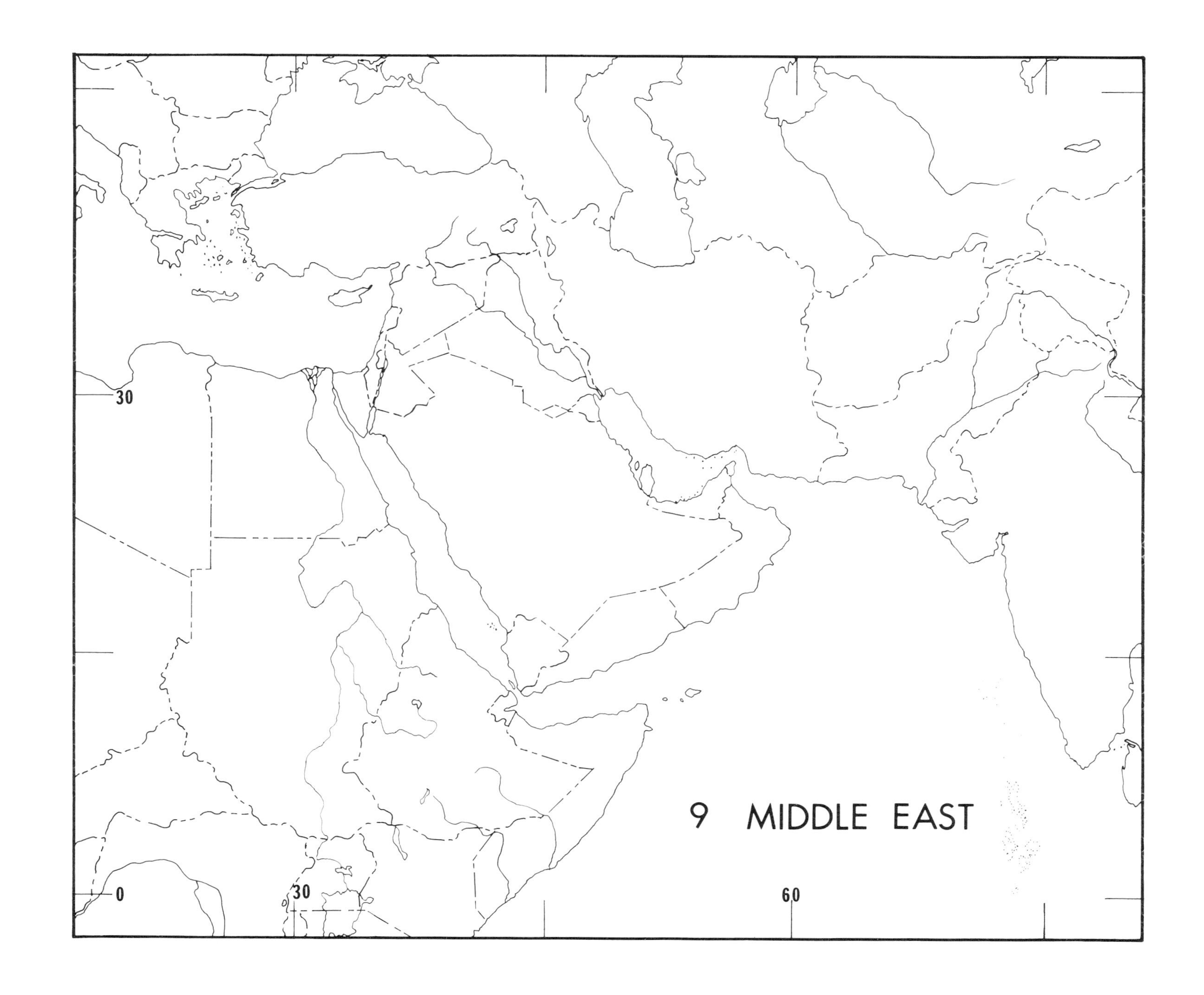
9 MIDDLE EAST
30
30
0
60

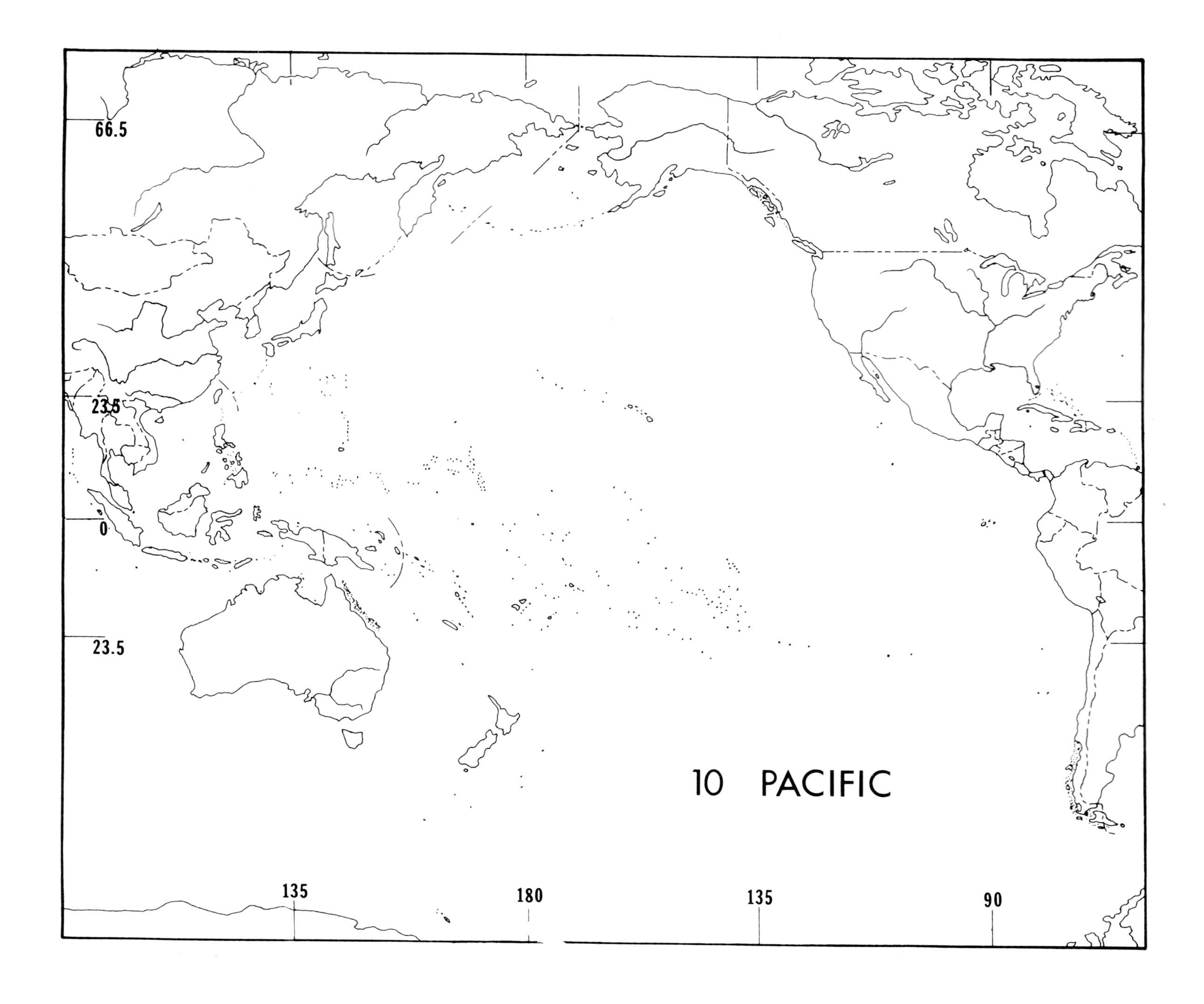
66.5
23.5
0
23.5
10 PACIFIC
135
180
135
90